D1134604

EXPLORING SCIENCE IN MUSEUMS

New Research in Museum Studies: An International Series

Series Editor Professor Susan Pearce (Director, Department of Museum Studies, University of Leicester)
Reviews Editor Dr Eilean Hooper-Greenhill

This important series is designed to act as a forum for the dissemination and discussion of new research currently being undertaken in the field of museum studies. It covers the whole museum field and, broadly, addresses the history and operation of the museum as a cultural phenomenon. The papers are of a high academic standard, but they are also intended to relate directly to matters of immediate museum concern.

NEW RESEARCH IN MUSEUM STUDIES

An International Series

6

Exploring Science in Museums

Edited by Susan Pearce

ATHLONE
London & Atlantic Highlands, NJ

First published 1996 by The Athlone Press
1 Park Drive, London NW11 7SG and
165 First Avenue, Atlantic Highlands, NJ 07716

British Library Cataloguing in Publication Data
A catalogue record for this book is available from the British Library

ISBN 0 485 90006 8

Library of Congress Cataloging-in-Publication Data
Exploring science in museums / edited by Susan M. Pearce.
p. cm. — (New research in museum studies: vol. 6)
Includes bibliographical references and index.
ISBN 0–485–90006–8
1. Museums—Educational aspects. 2. Science museums—Educational aspects. I. Pearce, Susan M. II. Series: New research in museum studies: 6.
Q105.A1E97 1996
507′.5—dc20 95—30352
CIP

Typeset by WestKey Ltd, Falmouth, Cornwall
Printed and bound in Great Britain by Bookcraft (Bath) Ltd.

Contents

List of Figures

Editorial Introduction

SUSAN M. PEARCE

Exploring Science in Museums is a title chosen deliberately for its ambiguity. It can mean either 'the exploration in museums of the fundamental approaches and attitudes which make up what is seen to be the absolute principle of scientific knowledge' or 'the exploration in museums of the notion of science as an activity with outcomes'. The phrase 'in museums' is a key element in both these descriptions, because the view taken of museums, collections and exhibitions, and their role in the scheme of things, is a touchstone for the power and interest of the informing scientific vision.

We may express the difference along these lines. If scientific knowledge is seen as absolute, and therefore as pure both of past and of human contact, then the relationship between it and museums can be expressed succinctly as:

museum	:	science
collections	:	ideas
matter	:	mind
past	:	future
graveyard	:	vital arena
trophies	:	hunters
passive exhibition	:	active seeking
treasuries	:	wealth-creators
lay people	:	knowledge-owning scientists.

Here museums and their collections are the outworn and discarded chrysalis, left to decay while the butterfly perpetually takes flight. The very best that museums can do is to forget the dead weight of their collections and concentrate upon exhibitions which will, at any rate, translate arcane ideas into layman's language, and so help to encourage a climate of opinion sympathetic to science and all its (expensive) works. It will be noticed that this scenario shares the view that the material world is inevitably passive and inert, without animating force of its own to influence events, a view rooted deep in the mind-set of the western tradition. Similarly, the role of science within it is god-like, enjoying a divine independance from contingency, time and circumstance.

The alternative view, which sees science as an activity with outcomes, would express the relationship between science and museums very differently:

museum	:	science
collections	:	inspiration
tangible evidence	:	constructed theory
history	:	politics
culture	:	process
informing exhibition	:	public awareness.

Here, science is seen to be a part of human culture and the kinds of understanding which it yields to be an organic growth from its past activities and its present political position. Accordingly, museums are not the dead past but the active, inspirational present, the living culture (in both the humanist and the scientific sense) which scientists must understand because it accounts for their present and will influence how they can create their futures.

It is perspectives upon this central problem which the papers in this volume address. Doughty and Knell, through the study of museum geology – a characteristic subject area – show how the political and cultural history of the discipline has affected the ways in which museums have been regarded. The theme is taken up by Arnold who pursues it in relation to science as a whole. Simmons presents a critique of 'hands-on' science centres in Britain, and Arthur Molella and Carlene Stephens discuss the highly controversial 'Science in American Life' exhibition at the Smithsonian's National Museum of American History, which draws politics into the centre of the frame. The problems inherent in interactive museum displays are pursued by Gillian Thomas and Tim Coulton for Britain, and Ibrahim Yahya for India. John Durant's contribution draws these themes together in a provocative way.

If a single message comes from these papers, it is that ambiguity and ambivalence are always with us, in 'exploring' science as in everything else. Museum involvement in science may be controversial; but it can be a healthy controversy (as Kaushik's review suggests) and a fruitful tension.

PART ONE

1

Museums and Geology

P. S. DOUGHTY

A GRAND VISION

There was a time when the practice of geology in museums was directly, indeed inseparably, linked to the wider practical application of the science. On 12 May 1851 His Royal Highness the Prince Consort opened to the public the Museum of Practical Geology in the presence of the élite of London society, ambassadors, peers of the realm, members of parliament and a select group of the nation's most prominent scientists. The Museum was the brainchild of Sir Henry Thomas De la Beche, Director of the Geological Survey of Great Britain and Ireland, who made the preliminary address. It was followed by that of Prince Albert, present not merely to decorate and ennoble the proceedings, but to add his considerable understanding and support to scientific research and the dissemination of the geological concepts promoted by De la Beche, somewhat endangered by the death in 1850 of his faithful champion, the former prime minister, Sir Robert Peel. In his speech the Prince Consort made reference to the ever increasing interest afforded science, and the unerring rise of specialization within it leading to useful and practical applications. He continued:

> In this view it is impossible to estimate too highly the advantages to be derived from an institution like this, intended to direct the researches of science, and to apply their results to the development of the immense mineral riches granted by the bounty of Providence to our isles, and their numerous colonial dependencies.

The agenda for this museum was thereby established from its inception with a clarity that few others attain, the result of the deliberations of a group of highly focused and politically influential geologists (Flett, 1937).

The Museum, and De la Beche's concept, were innovative

beyond contemporary belief and carried with them a political conviction and professional swagger that faded only with the appointment of the fourth Director, Sir Archibald Geikie, in 1882. The building itself occupied a prime site at 28 Jermyn Street, with a spanking new frontage in the Italian palazzo style at 221 Piccadilly. The stone for the facings was selected by De la Beche himself, Anston dolomite from a site near Macclesfield, the selfsame stone he had earlier selected for the new Houses of Parliament at Westminster. Inside was a great public auditorium illuminated behind the lectern and demonstration bench by the arcade of six massive windows of the Piccadilly frontage. Upstairs and extending into the building beyond was the exhibition hall itself, surmounted by two balcony galleries, all packed with their carefully selected and arranged exhibits. In the day, light poured into the space through the huge roof lights glazed with the massive curved panes then in fashionable use in ferro-vitreous structures like the Crystal Palace. Huge gas globes suspended from the roof provided winter and night-time illumination. The Museum was also furnished with public laboratories, a library, and a fine map collection.

It was the first government edifice entirely devoted to the study of science and was a fully rounded and generous concept. Four separate services were to be provided, each by its own agency: the Geological Survey, the Museum of Practical Geology, the School of Mines, and the Mining Record Office. There were four separate chairs, of Chemistry, Zoology, Mining, and Geology, and because all were under the Directorial supervision of De la Beche there was active encouragement for the staff to meet, mix and share their expertise. And what a staff they were, a galaxy of the finest talent then available. The Professor of Chemistry was Lyon Playfair, destined to be Lord Playfair of St Andrews; the zoologist was Edward Forbes, formerly palaeontologist to the Geological Survey and considered by many to be the outstanding intellect. The first Professor of Mining was Sir Warrington Smyth, and Andrew Crombie Ramsay was the first occupant of the Chair of Geology.

The combination of an active field survey and a school of science as an integral part of a *practical* museum, and it was the museum that was the central focus, created a dynamo to drive public science policy. There was a fusion of academic sciences and scientific practice that obviated the need to argue the role and significance of museums because all the staff in all four of the agencies shared and collaborated in the common quest.

In 1854 the Museum, the School of Mines, and the Geological Survey were transferred to the newly formed Department of Science and Art. Some changes occurred which evidently displeased some of the staff, and the by now seriously ill De la Beche. There were suggestions, strongly opposed by the Professor of Mining and Mineralogy and by John Percy, the new metallurgist, that the School of Mines should be established separately. Both insisted that the connection with the Museum and Survey was essential to their work. It is difficult now to read the politics behind the events, particularly since the public stature and esteem of Jermyn Street were at their zenith. The course of Evening Lectures to Working Men offered in 1853, and taught by a wide range of the staff, had its full allocation of 600 places filled in just two days. The year 1854 saw the arrival of Thomas Henry Huxley to the staff, an irresistible lecturer, soon to become a powerful intellectual force in support of Darwin's evolutionary theory. The death of De la Beche in 1855 in the midst of the debate removed the source and energy driving the vision. Laboratories and the mining teaching staff, with few exceptions, moved to South Kensington, and from 1862 the new formal title of the Royal School of Mines began to appear in association with the new venue.

Jermyn Street retained its vigour and prominence into the early 1880s, but the appointment of Sir Archibald Geikie to the Directorship of the Geological Survey in 1882 saw a major shift in priorities. Gradually the grand vision that was De la Beche's faded in an increasingly specialist Survey focused on or, depending on the view taken, preoccupied by, the production of maps. As staff resigned, moved or died, their posts were not filled. Chemical and metallurgical laboratories became work spaces for the general Survey. From 1883, as the books were transferred to South Kensington, the last faint glimmer of the great ideal was snuffed out. Integrated grand science was no more.

The Museum continued, but very much as a conventional institution, providing a useful level of public contact, largely resulting from the calibre of the Survey staff. In the First World War the foundations appear to have been damaged by a bomb, and remedial work closed the building in the early 1930s, a loss passing almost without note. Yearly attendances by this time were less than 25,000.

The 1927 Royal Commission on Government Museums and Galleries recognized a hopeless situation, and a delegation of its members made representations to the Chancellor of the Exchequer, accepted by Parliament in December of the same year, to move the Museum to a new site in South Kensington.

The new building was commenced in 1929 and although completed in 1933 was not handed over to Survey staff until 1934. The ceremonial opening on 3 July 1935 by the Duke of York revealed an absolutely standard and bland interior with pedantic presentations, largely the work of the Survey staff rather than the curators. It was, indeed, an arm of Survey, though by now a somewhat withered limb, with none of the power and flexion of De la Beche's master construct. To successive Directors of the Geological Survey the Museum appeared less and less relevant to what they perceived as their central role. The 1935 exhibits remained on the upper floors into the 1990s. In the early 1980s despite some vigorous new displays on the ground floor and first floor balcony, the Museum was regarded as such an encumbrance by the Survey that it was passed to the control of the British Museum (Natural History). The final, perhaps unwitting, act of betrayal of the noble concept of 1851 was committed as the Museum was dissociated from the last of the three founding forces to which it had been harnessed.

THE NEW ORDER

To some extent before De la Beche's 1851 coup, and to a much greater extent for the rest of the century, public museums grew up around the kingdom. Although no other geological museum was founded, many of this new generation had geological collections. The generalist nature of curators was the norm in the nineteenth century and for the next half century, but from around 1950 onwards, a swelling stream of graduates was drawn to curation and an inexorable change began.

Throughout the 1960s and the early 70s the small band of UK geological curators became increasingly aware of each other from accidental, coincidental and consequently unstructured encounters. Such meetings served to emphasize their professional isolation and disarray and to fuel an unease. Curation received no official acknowledgement in the main institutions of geology such as the Geological Society of London, the Palaeontological Association, and the Geologists' Association. There was an urgent need to be able to organize museum geologists into a body capable of effective professional action, but the most obvious umbrella organization to provide the impetus, the Museums Association, appeared unwilling to grant the requisite status to what was regarded as a special interest group. In the early 1970s a growing awareness of threats to collections added a sense of urgency, and it was the Leicester Museum and Art Gallery Earth Sciences staff who finally grasped the nettle and called a meeting of all interested parties. Preliminary discussions were conducted with the Geological Society of London and finally in May 1974 the Geological Curators' Group (GCG) came into existence, affiliated to the Geological Society. Its subsequent impact on geology in UK museums revolutionized curatorial practice and perceptions of museum professionalism.

Initially its intentions were modest. There could be no grand design to match De la Beche's, because there was no unitary control, but more pressingly there was no grasp of the scale, scope and condition of geological collections in the UK, indeed no appreciation even of the number of active geological curators practising in the field. Necessarily the first constitution of the Group set itself the fundamental objectives of provision of curatorial advice, the surveillance of geological collections with a view to safeguarding their well-being, the preparation of a code of practice for the curation and deployment of collections – a unique commitment at that time – and the initiation of a survey of geological collections and curation to establish some kind of baseline for future activity. Remarkably for such a small body, with no income beyond its own subscriptions, these objectives were substantially achieved in the first decade (Doughty, 1985).

So it was that in 1984 GCG felt the need to examine the wide range of issues then facing museum geology with a view to establishing a new set of objectives to carry forward in its second decade. The debate included documentation standards, collection indexes, curatorial development by means of specially devised activities, directories of specialist curators, and underdeveloped areas of geological activity such as mineralogy, conservation and geological technology. Collection issues remained as the overriding concerns but there was a growing realization that resourcing to carry through even the most basic projects required more than museum support: political awareness was essential and the motivating force stimulating political activity was public interest and concern. The major problem was immediately identified as the lack of any significant public profile for geology as a science, and the inability of the small handful of museum geologists, perhaps around 80 in number excluding the staff in the Natural History Museum in London, to change the situation unaided. It would require the mobilization of geologists in all parts of professional geology to significantly change the public perception of the science, and it was in this context that the professional credibility of museum geologists in their wider professional group became an issue. How did geological curators relate to their colleagues in the wider world of academia, public agency, industry and commerce?

Geology, as with all museum-based science in the twentieth century, found itself at a structural disadvantage, isolated from mainstream public science within the Department of Education and Science. With the exception of the Geological Museum and the Natural History Museum, all other museums (apart from the service ones) were either funded through the Office of Arts and Libraries (OAL), the case for national museums, or by local authorities through committees, generally of the leisure and amenity type. There appeared to be no national policy for the former, certainly no recognizable written objectives, and the latter were subject to purely local agendas but again largely unstructured. Responsibility for museums at national level was the charge of a minister for the arts, and without exception this brief had been held by politicians with arts interests. This peculiarly British concept of culture placed science in a peripheral position even within museums, despite the occasional protestation of the OAL to the contrary. The contrast with De la Beche's vision of the role of museum science, and particularly of geological science, within the economy and the cultural and professional life of the country could hardly been more stark.

A PROFESSION ALOOF

The other major barrier to public awareness was the attitude of the science of geology itself as manifested by industry, the agencies and the universities. Up to the early 1980s museums were yoked to a science which had no public perception, and appeared to seek none, a situation deriving from an affectation of immunity to public opinion.

It was a profession studiously ignoring the cardinal rule of tutelage, that a science that cannot address the values, needs and questioning of society, and inspire its support, will lose its public patronage (Slotten, 1994). It shared a presumption common to many utilitarian sciences at that time, based on a somewhat myopic stasis. The argument proceeded from the axiom that geology was essential to the wealth of the nation: it underpinned the entire industrial process and was uniquely responsible for the recognition and location of mineral raw materials and aggregates. It also identified the primary location of all the fuels utilized in the generation of power, to refine the raw materials and drive the major manufacturies. Since the entire developed world and its trade, outside foodstuffs, was conducted in these materials, their refined derivatives and the resulting range of technological products, there would always be a demand for geologists. This hubris was reflected in higher education where universities tailored course content to provide products able to sustain this construct: graduates to provide the expertise demanded by the basic mining and quarrying conglomerates and the agencies supporting them. This was a closed professional community, shunning public opinion. It also carried a slight edge of intellectual arrogance. Despite the centrality of British geologists to the fundamental concepts of geology, involving giants such as James Hutton and Charles Lyell; despite their dominance in the plotting of geological time, particularly associated with Adam Sedgwick and Roderick Murchison; despite their creation of modern geological cartography, the unique contribution of William Smith; and despite the elucidation of organic evolution by Charles Darwin and Alfred Russell Wallace, there was no interest in establishing a pride in their achievement and their place in cultural history.

There was also a view, quite widely articulated, that the sophistication of late twentieth-century earth sciences was such that to explain it to a general audience would trivialize its substance. Virtually no popular literature was available and the few authors who ventured into the field were often viewed quizzically and condemned as vulgar populists. All this at a time when one of the greatest scientific syntheses of the twentieth century, the concept of plate tectonics, was emerging in all its grandeur on the macro scale, and all its intricacy of process. Nothing less than the machinery driving the planet was being explored and explained piece by wonderful piece. In public awareness terms it might not have been happening. This indifference rendered such an introverted professional community an unlikely and unpromising ally to curators in restoring public interest in science generally, and geology in particular. But a sequence of global economic events was about to shake these apparently impregnable foundations and expose the basic fallacies of geo-economic indispensability.

THE BUBBLE BURST

The 1970s witnessed the flexing of power wielded by the most durable of the century's major cartels, the Organisation of Petroleum Exporting Countries (OPEC). In 1973 and again in 1979 OPEC reached agreements to vastly increase the world price of crude oil and to regulate production to sustain the new price. Massive disruption of the global economy followed on both occasions, one result of which was to stimulate exploration programmes for new oil deposits worldwide in areas outside OPEC influence, in turn stimulating the demand for earth scientists. But in 1982 as these new oil and gas fields came on stream, some OPEC members broke quotas, forcing price reductions, and in a relatively short time there was a glut of cheap oil and gas. Competition in the world fuel markets brought all prices down, including that of coal, and the slow political strangulation of the British coal industry inevitably followed. The effect was further reinforced by a general fall in the price of the major industrial minerals as the commodity markets reacted to a steadily intensifying recession.

The net effect on the earth sciences community in the UK was severe. As raw materials costs fell, so investment dried up. With weak demand and an uncertain future UK fuel and mineral exploration and production companies, always major world players, began to wind down exploration programmes. Employment opportunities declined steeply and became fitful at exactly the time when large numbers of geologists with considerable experience were released onto the labour market as contracts ended or were bought out.

These developments had an immediate effect on industry's demand for geology graduates, and whether it was this or the claimed fall in the international status of UK geological higher education that forced the next major change is difficult to elucidate. In 1986 Professor E.R. Oxburgh was appointed to head a team charged by the University Grants Committee to examine the teaching of earth sciences in the universities. In 1987 they reported (Oxburgh, 1987), proposing the most fundamental change to the structures of higher education in the history of geology teaching in Britain. Under the pretext of remodelling to re-establish Britain with the international pack leaders, the report proposed a division of university departments into three tiers. Level 1 departments would be centres of excellence, i.e. large, well-equipped in research terms, and around 12 in number. Level 2 departments would be smaller, again around 12 in number, focused on teaching courses to M.Sc. level, and with a far lower research profile and no expensive equipment. Level 3 departments would simply provide earth science support teaching as a component of other courses. The response within university departments was something akin to panic, stimulating a great deal of committee activity before a revised scheme was agreed (*Geology Today*, 1987a, b; 1988a, b).

During this period it became evident that a large number of well-established departments would become either interdisciplinary in nature, or provide joint honours teaching only, and that a handful would effectively close completely. In the haste to establish teaching and research status, the collections of the closing departments, or those moving out of mainstream geology, were almost completely overlooked. In a closely argued and withering attack on the inadequacy of the debate on significant collections, Taylor (1988; 1990a, b) focused an unblinking stare on the fate of key collections, particularly the principles underlying relocation and the funding of curation and conservation. Such forceful intervention led to the Universities Funding Council supporting the development of new museum storage facilities with varying degrees of curatorial cover in the university geology departments in Birmingham, Cambridge, Glasgow, Manchester and Oxford. In university terms, curation had come of age.

As subsequent developments have shown, the only lasting element of Oxburgh's recommendations has been the development of a handful of large, reasonably well equipped, departments. Perhaps this was the intention all along. The tragedy has been the closure of some well-established departments at exactly the time that government was contemplating the conversion of polytechnics to universities, creating dozens of new earth science courses and hundreds of additional graduates.

Contemporaneous with the industrial and academic change, the chief agency of earth sciences, the Institute of Geological Sciences (IGS) as the Survey had become, embarked on a period of evolutionary, almost extinctive, change. The Thatcher Governments of the 1980s progressively confined the funding base of the research councils. Agencies were encouraged to generate a proportion of their funding. Internal charging between the divisions of agencies became an unwelcome feature of professional life, and short-termism created major aberrations. Where units could not generate immediate income they were liable to drastic pruning. So it was that the Palaeontology Unit of IGS, with responsibility for an estimated 7 to 11 million fossils, the fundamental collection of British palaeontology, was reduced from 40 staff, to a small rump, predominantly of micropalaeontolgists. Market forces operated to the point where almost any work within the capacity of IGS to perform was pursued, leading to the position where in 1978–79 it was able to generate more than 80 per cent of its income (Wilson, 1985). This wholesale enthusiasm for free market commercialism alarmed large sections of the scientific community who began to question the role of IGS. Inevitably the unpredictable nature of the market led to piecemeal project work, and any sense of direction was lost. The central work of survey, and mapping, declined steeply just at the time when the new-found earning capacity of the IGS was reflected in a constriction of grant funding. A further complication quickly appeared in the form of a proliferation of specialist consultancies established by the geologists shaken out by industry. Squeezed between impoverished research councils and bullish consultancies, IGS found income generation increasing difficult, and having so readily abandoned mapping it became difficult credibly to re-establish basic survey work and the traditional provision of geological information. The climb-back is slow and now further complicated by contracting out, easily imposed against the recent historic background.

In museological terms the changes in IGS have weakened its public defences and damaged both access to professional expertise, and to the national collections, removed to a village in Nottinghamshire. It divested itself of the Geological Museum in this period, and lost its macropalaeontological team, a unique resource which cannot be rebuilt on demand. It retains its seminal and massive British collection, with little more than a fig-leaf of curatorial cover, and an even smaller proportion of its specimen data in machine-readable form.

Geology in the UK suddenly found itself on the lip of the abyss, confined on all sides except professional free fall. The scientific press, the science research councils, the Universities Funding Council, and the British Association for the Advancement of Science, all in their various ways turned to government expecting political redress for their plight. The basic truth was quite simply that the Thatcher Governments of the 1980s had never valued science. While there was a buoyant economy supporting a replete electorate there was no sound political reason for serious and expensive investment in primary research. The market was performing properly and in any case the national economy was increasingly running on service industries while the manufacturing base was being permitted to contract rapidly and, as many critics felt, alarmingly.

Geology desperately needed political support, but the driving force motivating politicians, a wide, enthusiastic and resolute public lobby (Slotten, 1994), was conspicuously wanting. The years of disinterest, of almost smug satisfaction based on the arguments of impregnable geo-economics, were now exposed for what they had always been, a confined and blinkered view of a much grander and culturally enriching subject crying out for public expression. The earth science establishment began to cast around for foundations on which to build a public base, and suddenly museums were recognized as having a wider role than was ever conceded in the 70s and early 80s.

SECURED FOUNDATIONS

In the longer term the reverses of the 1980s may provide the germinal tissue of a much more aware, balanced and robust science. It is still too early to judge. But there were a few interesting growth areas in geology from 1978 and throughout the turmoil of the 1980s. The most notable was geological conservation. The Geological Conservation Review of the Nature Conservancy was a major undertaking, identifying and scheduling over 2,200 of the principal heritage sites in all parts of geology and all areas of Britain (NCC, 1990). It also underpinned the great project then scheduling the rapidly dwindling areas of biological interest, although at the time it was not always evident that the two were complementary. That this was the case now emerges in English Nature's natural areas strategy, an enormously interesting concept which is only possible because the life and earth science surveys proceeded in tandem. Examination of the natural areas maps shows that they could almost be read as geological maps, so closely does the biology relate to its rock foundations.

Momentum on these large-scale projects was maintained because environmental issues became a major public concern in the late 1970s and the evident political impact of the initially small green parties led to rapid grafting of environment issues into the agendas of the main parties. Importantly, an undeniable green fringe began to suffuse the body of blue government policy.

Museums have played an active role in this movement. When the Geological Curators' Group sought affiliation to the Geological Society of London, one of the terms of the agreement was that the Group would undertake the creation and development of the national Geological Site Recording Scheme (Cooper *et al.* 1980). The acceptance by the Group of a major national project unrelated to geological collections and curation was a source of unease amongst some of its members. By the late 1980s after 15 years of determined effort, this apparently risky project had become accepted as a fundamental aspect of museum work. As the implications of the scheme began to bite it was recognized that the role of museums was not simply that of repository for the major materials of geology, it was also provision of information about them, and about the sites from which they originated. Much research on museum collections poses questions that can only be answered by detailed investigation of the site from which the specimens originate. In the information revolution science curators early recognized their unique position as prime holders of principal databases of material culture and heritage. Site documentation is a direct and logical extension of that role.

NEW ALLIANCES

The advantages resulting from the turbulence of the 1980s were largely to the museums in the regions of the UK. The British Museum (Natural History) in London, and IGS's Geological Museum before amalgamation had always had wider geological contacts, and they continued. Up to the 1980s it was virtually unknown for the great bulk of museums to have any kind of association with the main body of the geological profession. The nature of the change in these museums was evolutionary, one of progressive exploration by the users followed by consolidation. It would be wrong to think that the bodies concerned were changed in any radical way by their increasing use of museum resources, but the expanding practice showed the acceptance afforded to museums and a respect for their professionalism; a recognition of the work of the first fully graduate generation of curators.

Mineral exploration companies make particular demands. Collections are increasingly consulted, particularly specimens key to particular problems: anything from individual fossils, coal or oil samples, to ore minerals and vital rock types. Precise biostratigraphy (the dating of rocks from their contained fossils) is an essential element in much economic geology and in preliminary investigations the availability of collections of fossils with known biostratigraphic significance becomes important. The variety of such fossils is wide, but much practice is now based on microfossils. Curators from 1950s onward broadened the collecting policies of institutions, and collection increasingly included such material as foraminifera, plant spores, ostracodes, acritarchs and conodonts. There is a wide spectrum of exploration techniques available to geologists now, from remote sensing to fine image seismography, and existing technologies are improved and new ones added steadily. But at some stage the electronic and paper-trace potential of a site has to be tested on the ground. Many companies now consult museum material early in this process as a quick and convenient way to check what is already known and to conduct preliminary appraisals. Almost all companies admit that an element of their exploration programmes is the revisiting of sites where material of economic interest has been known to occur previously. This gives special value to historic economic mineral specimens, particularly when they are well localized. Many museums, or amateurs and volunteers associated with them, have taken particular interest in old mine sites, even very limited occurrences of ore minerals, as structured regional research projects. Museums also accumulate historic maps, often annotated or with mineral symbols, historic manuscript reports, diaries, commonplace books, mining prospectuses for long-vanished companies, indeed archive materials of all kinds which relate to their collecting interests. Plans of old and abandoned mine workings, rare monographs, early memoirs, historic photographs, and a range of other illustrative materials from sketches to engravings are the stock-in-trade of such museum archives and may save mineral companies much effort and expenditure. There are cases where collections have substantially remodelled exploration programmes.

The considerable effort devoted to inspecting and listing the built heritage of the UK has revived interest in building stones, both to accurately describe the fabric of buildings, and to identify sources for current or future repair. This has led to a steady stream of architects and building conservators exploring collections and expertise. The principal collections of UK building and decorative stones are in the Sedgwick Museum, Cambridge (Watson, 1911, 1916), and the Natural History Museum, London (removed from the Geological Museum), but other collections dotted around the country together constitute a considerable resource. Enquiries have risen to the point where a national database to match stones, or to suggest appropriate substitutes where the original is no longer available, is in active discussion.

Some of the UK's most prominent gemmologists are curators of the small numbers of richly diverse gemstone collections, and museum gemmologists offer gem identification as part of their service. Collections are frequently used for comparison, instruction and research. Many commercial jewellers trading remotely from the main gemmology laboratories are discovering unsuspected local expertise.

A very large number of museums now have immense numbers of specimen records in machine readable form and large numbers of site and historical records in databases. Internet, SuperJanet and similar national and international communications media are elevating the status of these kinds of data in ways beyond the original and optimistic expectations of the visionary curators who were early into the field. Industry is already exploiting this potential.

Extractive industries are meeting increasing problems in obtaining planning permission and public acceptability for their work and they have begun to realize some less obvious potential offered by museums with a strong geological presence. By using such museums as repositories for their materials, and as record centres for their sites, they are taking the first steps towards a more conciliatory view of an environmentally aware public. This development has risks for museums, but handled with care and professional integrity it has the potential to contribute intelligently to the debate on consumerism and exploitation of natural resources. Many companies operating at a distance from their head offices use museums able to provide quality venues and services for local meetings and events. Whilst the venue may often be almost accidental, it offers potential for first contacts or the development of relationships. Some companies admit that museums have been their entrée into local geological communities.

Another aspect of industry is worthy of museum consideration. The major geological industries take pride in their achievements and have a view of themselves, or at least an image of their development, that they wish to promote. They have an abiding interest in their early development, in people, events, technologies and their company's economic importance. This is usually supported by an unofficial and dispersed archive, quite often with some kind of historian. Such histories are legitimate and productive subject matter for museums but the museum must retain editorial control.

The relationship between museums and industry is relatively new, and both sides are only now starting to learn how to develop it to mutual advantage. The expansion of these contacts has emerged from the maturing of curatorial expertise at a time when industry has needed a shop window. It is likely that contacts and collaborations will expand.

The explosive growth of environmental consultancies in the last decade has been a natural development from political concern for the environment at a time of political distaste for the expansion of public services. Museums have moved from a position where consultants were almost unknown to one where they make regular and expanding use of resources.

All major planning decisions now require environmental impact assessments and in the case of sensitive landscapes may extend to public hearings. Consultants need the longest-term environmental histories of sites that they can find, and these invariably lead to the long established museums with natural history departments. Most consultants are from biological and environmental sciences backgrounds which means that they need particular help in making judgements on geological issues. Although they must consult widely, they appear to prefer museum opinion and sources for good reasons. Museums are accustomed to thinking in special site terms, they are used to communicating scientific information to non-specialist, non-technical users and they invariably hold the relevant databases with strong historical components and an emphasis on unique events and occurrences. Add to this the broad environmental view, including not just the geology, but biology, archaeology, history and some knowledge of architecture, records of hazard events such as land-slips, floods, rock-falls, bog-bursts and pollution events, and it is evident why the museum has such appeal. In addition to the unique skills specific to the curator and intrinsic to collections, museums hold relevant mineral and planning information, specialist maps, associated academic literature and all the standard office facilities of the information age. The consultant/curator relationship has brought museums nearer centre stage in the entire environment planning process and it is likely to evolve much further.

Museum relationships with the universities and higher education establishments generally were very weak until comparatively recently. When GCG was established, university staff formed only a very small proportion of the membership. That proportion has risen over the years so that staff in every major university museum are now active members. The general activity within the Group has made university curators aware of the wealth of geological material, experience and expertise available across the country, and local and national museum curators more conscious of the priorities in establishments of higher education. The relationship has taken on completely new directions, partly as a consequence of the implementation of the Oxburgh Report (Oxburgh, 1987), partly as a result of curators actively researching their own collections and seeking academic opinion, which in turn has led academics to a growing familiarity with museum material, and partly due to the straitened circumstances in which universities now find themselves. The reaction to Oxburgh made universities aware of an informed and organized group of museum opinion, concerned about curatorial standards and the fate of important collections. The significance of repository and the obligations it imposed were uncompromisingly stated and could not readily be evaded. That concern was already being championed by editors of many journals. They frequently require authors publishing material-based research to state clearly the repository in which the status specimens are deposited. The general academic community now accepts that the continuity of evidence in the history of ideas requires the fastidious application of curatorial standards. University researchers almost routinely look to museums early in their projects to take their important material, provide more specimens, and locate collections of interest.

The Federation for Natural Sciences Collection Research (FENSCORE), operational now for a generation of biological and geological curators, has led to the publication of regional statements (Hancock and Pettitt, 1981; Davis and Brewer, 1986; Hartley *et al.*, 1987; Stace *et al.*, 1988; Walley, 1993; Bateman and McKenna, 1993) locating general collections, named collections, elements of dispersed collections, their composition and important components, the areas from which they originate, personalities associated with them and a variety of related information including type specimen indexes. A concise statement of museums with geological collections has recently appeared (Nudds, 1994) and may establish a new trend in presentation. Museums are increasingly partners in research, a direct consequence of grant requirements demanding collaboration as an essential condition. This new departure appears to be working well, shifting the axis of research towards programmes of museum interest for the first time.

Many of the second and third level university departments are under pressure from the increased student numbers the system now compels, and for many, even with the new levels, financial shortfalls are leading to depleted staffs incapable of meeting lecture commitments. A number of universities are solving the problem by inviting geologists outside their departments to teach elements of courses. Whilst the importance of this trend can be overstated, it is interesting to see curators lecturing, demonstrating and providing field instruction. The introduction of museum specimens into palaeontology and mineralogy courses gives many students their first experience of cabinet quality material and a new view of the relevance of museums to the practice of their profession in their future careers.

Further down the age range, at primary and secondary levels of teaching, the pace of change has been relentless, leaving the teaching profession bewildered and bemused at teaching commitments for which they have no training and in which they have no background. This has been particularly true in relation to the sciences, where the requirement is a response to the declining numbers of pupils with scientific interests in secondary and higher education, and the recognition that in an age dominated by science and technology, society should have a level of scientific literacy sufficient to permit its citizens to relate to a science-driven world. Ultimately a society that cannot uphold the values of science will not have the capacity to benefit from it (Slotten, 1994).

Museum education departments with scientific staff have been particularly well placed to assist both in the process of curriculum development and in the basic and refresher training of teachers seeking a scientific content for their courses. The focus in this work has not been the specialist geology teacher but the general teacher. Museum educators have been able to assist infrastructure in developing teaching materials, special school events, and links into practically useful organizations such as BAYS, the youth section of the British Association for the Advancement of Science which runs highly successful and enthusiastically supported programmes.

Even on their familiar ground, museums are able to respond to the needs of their employing authorities in ways that complement local aspirations. Museum geologists now find themselves involved not simply in site documentation, but in site protection and promotion, as local authorities assume cultural and environmental control for larger aspects of their regions. Tourism is developing rapidly in areas previously considered unsuitable or unpromising, particularly those blighted by heavy industry. As the manufacturing base has declined, and the air and water begin to clear, interesting landscapes emerge afresh, often revealing directly their founding raw materials. Interpretations of these former wastelands frequently begin from their foundations: quarries, mines, spoils, railway cuttings, natural escarpments, canal cuts, stone pits and open casts. Their potential is developed through local tourism offices and specialist consultants, but with curators playing a crucial and central function. Heritage and interpretation centres see curators as vital catalysts in brief preparation, in devising historic trails, designing theme tours, creating special exhibitions and activities, in leading re-explorations of town and city centres, rediscovering local building materials and introducing the new medley of decorative stones from around the world. They are uniquely qualified in this highly specialized background. Their local knowledge and expertise enables many authorities to stage unique and memorable events, exploiting such unpromising resources as abandoned mines, stone galleries and canal tunnels. Civic hospitality, incorporating such settings, imaginatively used, provide memorable elements of conference programmes, whilst contributing to the new and expanding network of scientific and industrial culture.

Even the bastions of government have shown interest in the potential of museum geology. Curators are drawn onto advisory bodies for government departments, usually providing landscape, geological conservation and cultural background. Museums are now commonly repositories for specimens collected as part of routine environmental surveys, and NERC (Natural Environment Research Council) and SERC (Science and Engineering Research Council) funded field research projects. The need to provide certain kinds of information to meet European legislative requirements and public demand for more open government, has led to the funding of museum-linked record centres initially on an experimental basis, but more recently for the longer term. Much of the information requirement in support of international agreements has required the commissioning of research to fill the voids, and for some of this work museum staff have been well positioned. Even within the British Geological Survey the pressure of opinion from GCG has resulted in joint seminars and workshops to examine documentation terminologies, especially a consistent approach to the plethora of lithostratigraphic names resulting from piecemeal mapping in the past. Museum staff have even been active players in the preparation of submissions of World Heritage Site proposals. In almost all these cases museum geologists have been consulted for their special expertise, arising directly from their curatorial experience. They have not had to stray beyond their legitimate activity to assist, and in most cases they are the most obvious sources of the specialist information sought.

Undoubtedly museums have been recognized as the evident shop windows of popular and not-so-popular earth science, and their success can be demonstrated particularly in the currently fashionable climate of performance measurement. Attendance figures are increasing, many attributable to special geological activities and events; there are more exhibitions, more popular publications, activity schemes such as 'Thumbs Up' for young people, more services offered and more users of established services. The one attempt made to create a popular nation-wide geological magazine, *Geology Today*, has relied heavily on the involvement of museum curators to provide interesting geology, described in plain language, and museum matters feature in an interesting and disproportionately large way.

Geology Today, however, has not made the major breakthrough essential to the creation of a strong and readily recognized profile for geology in popular periodical form, and in that respect it is symptomatic of many initiatives of many sciences. They have certainly achieved some improvement in public appeal but collectively they have not yet reached the critical mass needed to command a popular movement.

The last ten years have seen a recognition by all sciences that none alone can attract and hold a wide public audience. If government is to be persuaded that science is worthy of sustained support, it will only be because the whole edifice of science, together, confidently asserts its position in national life and culture. Science for every person must also have its place in the sun. There are stirrings. Science, Engineering and Technology weeks (SET 7 and SET 95) have received some support from the Minister. With better funding and a more consistent approach they have the potential to become great national science festivals. The will for the rebirth of a popular scientific culture is building, and museums are major players in this kind of arena.

BOLSTERING THE BULWARKS

The major institutions of science are the Royal Society and the British Association for the Advancement of Science (BAAS). The Royal Society is the senior scientific body, founded in 1645, and election to its fellowship is the highest accolade in UK science. Although great collectors, such as Sir Hans Sloane, are numbered among past presidents, it remains a body dedicated to the promotion of scientific research, and is influential in the development of national science policy. The Royal Society's interest in the public perception of science is reflected in COPUS (the Committee on the Public Understanding of Science) which it funds. COPUS has a wide-ranging brief with imaginative schemes in science education, popular representation of science and scientists, scientific lobbying, science in the arts, science information services, anti-science, scientific and parascientific belief systems and in science museum studies. Its media placement schemes have been particularly beneficial, and systems for combining elements of popular science are starting to emerge. Museums, or museum-based science centres, already perform many of these functions, and some museums are already examining radical remodelling to revolutionize science in museums for the millennium.

The British Association for the Advancement of Science was founded in 1831. Among the aims stated in its current corporate plan (BAAS, 1991) are;

- to increase awareness and understanding of science and technology among young people, key decision-makers and opinion-formers, and within the community at large.
- to achieve greater public recognition of the contributions that science and technology make to cultural and material well-being.

In furtherance of these aims it organizes a week-long annual meeting which is the flagship for the Association and by far the most important public scientific event on the national calendar. Typically several thousand delegates throng a university campus attending concurrent events of 16 sections. There is saturation press and broadcasting coverage and a palpable thrill of good science communicated to a general audience. The Geology Section was the third to form and has a long and distinguished history, and still provides popular and packed events in some years. The tone of the Section programme is set by its president, a position held for one year only. Continuity is provided by the recorder. The Section Committee in the last twenty years has been dominated by the universities to an unhealthy degree. The appointment of the president has almost amounted to passing the hat between university heads of department. There has not been a president from outside the universities and the Geological Survey in the last 20 years. An injection of exhibition and media interest in the form of three or four new committee members might yield a sensible balance when the public appeal of the more esoteric corners of geology are proposed as programme themes. Museums have been active in the Association in the last decade and the offer of special museum events and receptions is always gladly accepted. They are usually solidly supported and frequently oversubscribed. Museums have yet to devise a strategy for bodies like BAAS but if they choose to enter this forum they could be a reinvigorating force of great influence.

THROUGH THE MILLENNIUM

Museums in 1995 have clear and developing roles in the science of geology, a major shift from the position in 1974 when GCG took its first steps towards attempting integration across museums. The pendulum of involvement in the wider profession of geology has now swung so far that the problems confronting geological curators are those of selecting and defining the areas where they can intervene most effectively and where the best returns for individual museums and the museum movement lie. Different museums will see the opportunities differently, but the need for co-operation within geology, and of geology within the general ambit of science, is the obvious path to wider recognition and the realization of potential.

1985 saw the publication of *Guidelines for the Curation of Geological Materials* (Brunton *et al.*), a revolutionary publication based on the curatorial experience of around 20 curators. It had no peer in any other curatorial discipline and almost a decade passed before anything similar began to appear (MGC, 1992 onwards). It set the standards and defined the rigour of a new professionalism and it is that unique set of skills that is acknowledged and respected within the wider group of geologists. The period from 1985 also saw a driving energy to establish the specimen databases that are now coming strongly on stream, complemented by the FENSCORE collections catalogues. The history of UK museum geology, 1974 to date, has been to repair the omissions of the past and to arrive at a clear and widely available statement of what museums have, and where it is. Much of it is also secured in the medium term.

The challenge of the future, and the next mobilization of curatorial resources, will be towards the wider exploration and exploitation of collections and establishing their place in mainstream culture. The institutions of geology have moved too far for De la Beche's grand design ever to be recreated, but undoubtedly a new fusion based around museum resources is resurgent and, as the dust of the 1980's finally settles, the outlines of a new vision take on firmer form.

BIBLIOGRAPHY

BAAS (British Association for the Advancement of Science) (1991) *Challenge and Change: A Vision for the Future* (London: BAAS).

Bateman, J., and McKenna, G., (eds) (1993) 'Register of natural science collections in south east Britain', South Eastern Collections Research Unit and AMSSEE.

Brunton, C.H.C. *et al.*, (eds) (1985) 'Guidelines for the curation of geological materials', Geological Society Miscellaneous Paper no. 17, Geological Society of London.

Conservation Committee of the Geological Society (1984) *Record of the rocks* (London: Geological Society).

Cooper, J.A., Phillips, P.W., Sedman, K.W., and Stanley, M.F. (1980) *Geological Record Centre Handbook* (Cambridge: Museums Documentation Association).

Davis, P., and Brewer, C. (eds) (1986) 'A catalogue of natural science collections in north-east England', North of England Museum Service.

Doughty, P.S. (1985) 'The next ten years', *Geological Curator*, 4(1): 5–9.

Flett, J.S. (1937) *The first Hundred Years of the Geological Survey of Great Britain* (London: HMSO).

Geology Today (1987a) 'The Oxburgh Report', 3, (4): 112.

Geology Today (1987b) 'Oxburgh: The Report' and 'Oxburgh: The Debate', 3(5): 146–150.

Geology Today (1988a) 'Update on Oxburgh' and 'Oxburgh: opportunities missed', 4(2): 40–2.

Geology Today (1988b) 'Oxburgh: almost there?' 4(3): 78.

Hancock, E.G. and Pettitt, C.W. (eds) (1981) 'Register of natural science collections in north west England', Manchester Museum.

Hartley, M.M., Norris, A., Pettitt, C.W., Riley, T.H., and Stier, M.A. (eds) (1987) 'Register of natural science collections in Yorkshire and Humberside', Area Museum and Art Gallery Service for Yorkshire and Humberside, Leeds.

MGC (Museums and Galleries Commission) (1992–) *Standards in the Museum Care of . . .* (London: MCG).

NCC (Nature Conservancy Council) (1990) *Earth Science Conservation in Britain*: A Strategy (Peterborough: NCC).

Nudds, J.R. (ed.) (1994) 'Directory of British geological museums', Geological Society Miscellaneous Paper no. 18, London.

Oxburgh, E.R., Haggett, P., Jenkins, A.L., Muir Wood, A., Stewart, F., and Briden, J.C. (1987) *Strengthening University Earth Sciences* (London: University Grants Committee).

Slotten, H.R. (1994) *Patronage, Practice and the Culture of American Science* (Cambridge University Press).

Stace, H.E., Pettitt, C.W.A., and Waterston, C.D. (1988) *Natural Science Collections in Scotland* (Edinburgh: National Museums of Scotland).

Stanley, M.F. (1984) 'The record of the rocks', *Earth Science Conservation*, 21: 17–22.

Taylor, M.S. (1988) 'What will happen to the universities' geological collections in the post-Oxburgh world?', *Geology Today*, 4(4): 119–20.

Taylor, M.S. (1990a) 'The irrationalization of university geological collections', *Geology Today*, 6(1): 9–10.

Taylor, M.S. (1990b) 'The irrationalization of university geological collections – a postscript (no thanks to the UFC)', *Geology Today*, 6, (2): 54.

Watson, J. (1911) *British and Foreign Building Stones* (Cambridge University Press).

Watson, J. (1916) *Marbles and other Ornamental Stones* (Cambridge University Press).

Walley, G.P. (ed.) (1993) 'Register of natural science collections in the midlands of England', Nottingham City Museums.

Wilson, H.E. (1985) *Down to Earth: One Hundred and Fifty Years of the British Geological Survey* (Edinburgh and London: Scottish Academic Press).

2

The Roller-Coaster of Museum Geology

SIMON KNELL

INTRODUCTION

There is a strong perception that museums are primarily about history. This derives from the undoubted age of the majority of these institutions, but also the relatively recent growth of history-based museums and heritage attractions, including 'historic houses'. Even the 'science museum' is popularly perceived as an institution focused on the history of science and technology. These science museums communicate science but do not actively participate in it. Few sciences use collections as a scientific resource – perhaps only pathology, biology and geology.

Until the 1920s, the natural sciences dominated museum provision in Britain, and museums played an important part in the development of an understanding of this country's natural history. The establishment of the science of geology, once the most celebrated science in Britain, and largely a British invention, relied upon museums and collections. However, with the publication of the Doughty Report in 1981 that glorious past was revealed as having been betrayed; museums had failed in their most important responsibility (Doughty, 1981).

Doughty's report motivated operations to rescue lost collections, as well as a more general renaissance in museum geology: it is now better displayed, more professionally organized and increasingly popular with the public. But why hasn't this always been so? Is this simply a temporary respite? Is geology really more ammunition for 'declinists' to fire at government as they have done for more than 150 years (Morrell and Thackray, 1981: 47) or is its apparent fall the product of a more complex process?

THE RISE OF MUSEUM GEOLOGY

As geology entered into the nineteenth century it was just beginning to develop its own framework for scientific study. As the century progressed it became an increasingly powerful magnet for the cultured and wealthy classes, and for the existing scientific élite. It revealed a British landscape extraordinarily rich and diverse in geology, but largely unexplored. For the embryonic scientist/philosopher the opportunities for discovery were boundless, and the products of these discoveries remarkable.

By the 1820s, geology had worked its way into the public conscience; it occupied the conversations and correspondence of polite society. By 1840, geological meetings of the British Association for the Advancement of Science were attracting audiences of a thousand (Morrell and Thackray, 1984: 548; Rudwick, 1985: 250). Geology was the height of fashion – this was its 'Heroic Age'. The latest thinking seemed to change with every monthly magazine – almost as rapidly as the political scene. As a sign of sophistication, it was as essential for the aristocracy, gentry and growing middle classes to express a fascination for this new science as it was to possess a knowledge of the classics or European languages.

The popularity of geology in this period is not difficult to understand. These decades revealed a succession of remarkable discoveries including giant 'sea dragons', flying reptiles, extinct 'lizards' the size of the largest living mammal, and proof that exotic animals, such as hyenas, lived not long ago in Britain. For the provincial gentleman, works on regional geology by John Phillips, Gideon Mantell and others, provided models for imitation and a framework for local studies. The explosion of geological knowledge spawned palaeontological and geological syntheses, including elementary texts, which revealed rocks and fossils as the products of dynamic forces.

Geology opened doors not just to one new world but to many, and revealed facts more extraordinary and remarkable than anything created by contemporary fiction. Texts conveyed ideas which were both convincing and remarkable in elegant, and often romantic, prose. No special equipment or expertise was required to retrieve some useful fact; so unknown were local terrains that no qualifications were needed to philosophise on their products. As yet unadorned with the plethora of jargon terms with which the science would soon immerse itself, inductive and descriptive, geology was accessible to all. And unlike the abstract sciences of mathematics, astronomy, physics and chemistry, geology and natural history were tangible, comprehensible, romantic and everywhere. Geology and natural history permeated deeply into society and became normal activities among those who had sufficient wealth for leisure.

While the 'popular culture' of geology filled the local scene, the science's great dramas were played out by a small cast of eminent scientists who courted controversy and expounded facts with great eloquence. Their names became known to all – Mantell, Murchison, Buckland and others were mentioned in contemporary magazines with no need for introduction.

Morrell (1994: 312–13) suggests that the 'perpetual excitement' surrounding geology at this time was due to its economic benefits, its interest to the traveller, its adaptation to any scale of study and to its relationship with religion. The latter perhaps seems most surprising as it would appear that geological advance was certain to undermine literal interpretations of the Bible. Fundamentalist factions did snap at the heels of geological progress, and occasionally find their way into the debating chamber, but these were no more threatening than those which attempt to challenge the modern science (see, for example, Jones, 1980). The secular science disentangled itself from religion much later; for the church-going practitioners there was no conflict. The diaries, correspondence and contemporary literature expose a beneficial bond between the science and religion. William Buckland, probably the most influential geologist among the provincial gentry at this time, expresses this most overtly:

> When fully understood, it [geology] will be found a potent and consistent auxiliary to it [religion], exalting our conviction of the Power, and Wisdom, and Goodness of the Creator . . . No reasonable man can doubt that all the phenomena of the natural world derive their origins from God. (Buckland, 1836: 9)

In this optimistic scientific climate many existing museum collections were formed. Geological advance and opportunity, combined with widespread interest in natural history gave birth to the literary and philosophical movement which swept through much of Britain, but which was particularly strong in Yorkshire. Provincial Britain offered such unparalleled opportunities for geological research it seemed that every rock in every parish had the potential for turning this new science upside down. Unlike the private museums which could be found in most British towns and cities, the philosophers were first and foremost interested in using their collections to extend knowledge, as well as to facilitate self-improvement.

The Yorkshire Philosophical Society expressed a clear primary objective: illumination of the geology of the county. This became its focus both for research and illustration in the new museum. Aware of the private museums it succeeded, the Society articulated a mature understanding of the part collections might play in its objectives:

> It is one of the principal objects of the Society to encourage a taste for natural history; and it has always proposed to effect this, both by means of lectures on the various divisions of nature, and by collections in its museum, without which lectures of this description can scarcely be given. The value of such collections is not perhaps in general sufficiently understood; and the naturalist by whom they are formed is sometimes suspected of claiming the dignity of a science for pursuits little higher than the amusements of children. If the object of a collector be no more than to accumulate and to display, he is indeed very idly employed; but if his object be to acquire or to diffuse a more perfect knowledge of the works of creation, there cannot be a more rational or a more noble pursuit. To investigate the wisdom of nature, is an employment worthy of the most exalted understanding whether that wisdom be displayed in the configuration of a planet, or in the structure of a butterfly's wing. (Yorkshire Philosophical Society, 1828: 14)

Even in these early years, geology was very much an international research school with considerable cross-Channel and trans-Atlantic communication. In spite of this, the work of British geologists dominated the science's development. At its heart, the Geological Society provided the liveliest debating house in London (Rudwick, 1985: 18), and formed a museum which became the national repository for materials derived from geological advance. The British Museum purchased numerous important geological collections at this time but it played no active role until the appointment of Richard Owen in 1856.

For a few decades a geological research network spread throughout Britain. Established in the expanding urban centres, the provincial learned societies provided a focus for the intellectual pursuits of the growing middle classes – businessmen, medical men and clergy. They played a vital role in feeding local intelligence to the higher science; the doyens of the science would reward the local society with news of the latest research, collections or simply by association. The provincial philosophers followed research projects developed locally and those which emulated work undertaken elsewhere. The newly established British Association for the Advancement of Science formed a mobile focus for this research network.

The new and fashionable science of geology created the excitement required to keep a society enthralled; the society men eager to discover their own immortality built collections and provided reports which served the Heroic geologists. The interaction was both symbiotic and mutually catalytic. A framework for the investigation of British geology had evolved quite naturally, and privately funded museums were at its heart.

In their mode of operation the philosophical societies conformed to a pattern; they shared similar objectives and methods (Allen, 1976: 159). Their differences were primarily a reflection of local personalities and the size, and composition, of the social strata from which they were formed. They were an assertion of local status – 'so well calculated to promote the credit and advantage of the town, and the intellectual improvement of its inhabitants' (Whitby L and PS, 1826: 9). Relative status could be judged by the size of the museum building, the number of members or the society's links with the scientific élite. For those societies which could not compete on these terms a coup might be staged by acquiring more perfect and spectacular collections of fossils. 'Such collections, however, not only exhibit the natural productions of the province in which they are situated, but they may be taken as standards by which to gauge the scientific spirit of the neighbourhood' (Rudler, 1877: 17)[1].

In the county of York, the Yorkshire Philosophical Society reigned supreme. The societies of the smaller towns of Whitby and Scarborough, despite having, on their doorstep, some of the most accessible resources of fossils in Britain, were of a lower stratum. They struggled to maintain libraries or an interesting lecture programme. There simply were insufficient numbers of local philosophers in these small towns to provide the critical mass necessary for success on the scale of that seen in York. The York Society, through the numerous and influential contacts of William Venables Vernon Harcourt and the rapid scientific achievements of John

Phillips, became a crucial link in the development of British geology. York was a centre for science, and while valuable collections and expertise developed on the coast, these smaller and poorer societies often relied upon the York philosophers for intelligence and contacts with mainstream geology.

Increasingly powerful and assertive in its influence, the Yorkshire Museum became a particularly attractive repository for the finds of collectors, often in preference to the smaller, more local, society. John Dunn, fossil collector and secretary of the Scarborough Literary and Philosophical Society, feared the 'bubble reputation' of his society and preferred to donate important material to the rival society in York.[2] Placed in the York collections, there was a real chance that a specimen would be seen and used by science. If geological material was not seen by the monograph writers it was never going to reflect glory on the finder. Worse still, a rival's finds might be used instead!

Before John Phillips' (1829) explanation of the coastal geology of Yorkshire, the Scarborough collector had little difficulty in finding fossils which had not been described – the search for new species was the principle motivation for the coastal collectors. The publication of Phillips' book, far from dampening this urge to discover, gave the local philosophers new challenges to discover species or to prove error in those Phillips had described. By placing tempting morsels in the hands of a local society, the collector was not only more likely to have his material figured, but also to encourage eminent researchers to view his personal collections where more treasures might be revealed. If, like the Vicar of Wakefield, a collector was able to donate massive coal measure plants, six feet high and thirty inches in diameter, to all the museums in the region, publication was hardly necessary as the collector had built his own imposing monument.

The 'Heroic Age' intertwined provincial museums with pioneering scientific research in a way that was never to be seen again. They not only participated but also illustrated the advances both in static displays and in the lecture programme.

THE FALL OF MUSEUM GEOLOGY

By 1850 the real science of geology was becoming more rigorous and systematic, and its publications less approachable and more specialized. At the heart of this professionalization was Henry De La Beche's Geological Survey which was establishing new levels of resolution for data capture. De La Beche's government-funded Museum of Economic Geology increasingly adopted the role of repository for British geology. In a relatively short period of time the requirements of science became dissociated from the leisure interests of the populace. The once burgeoning philosophical societies and the museums they created were beginning to founder. Most had been dogged by severe financial difficulties throughout their existence. Particularly draining were the museums – often in magnificent and imposing buildings, packed to the gunwales with material, the care of which would take an army of curators.

The infectious excitement with which the museums had been built began to subside as the mainstays of the societies left or died. They had believed they were creating permanent institutions – 'not only in the present day but in future ages' (Whitby L and PS, 1826: 9) – but chose objectives which were achievable and therefore ephemeral. The Yorkshire Philosophical Society had, through the work of John Phillips, achieved its major objective by the mid 1830s. For a time the collections they amassed contained items of news, but soon they were news no longer; they were transformed instead into less evocative reference materials. Having fed on the excitement of contemporary science and their role within it, the Societies would have difficulty surviving without it. If geology was a catalyst to their development, they too contributed to the advance of the science to an extent that it soon outgrew its amateur roots.

Despite continued fears of revolution the philosophers of late Georgian Britain were also unable to predict the massive social changes which were to take place in the next quarter century. 'The tight grip of tradition had been largely broken, and that "ancient wisdom" in matters of belief, values and social relationships was being increasingly questioned' (Harrison, 1971: 144). The scientific world was no less politicized and mirrored the wider changes in society (Morrell and Thackray, 1981; MacLeod, 1983; Desmond, 1989). But while a more professional, and increasingly publicly funded, age dawned for geology, its heritage remained in the hands of amateur custodians in the 'private sector'.

Some philosophical societies survived, but most eventually gave over their museums to the local authorities now empowered to provide a museum service. 'Once removed from their originators, even the residual sense of purpose, identity, and obligation felt by the ailing societies was lost' (Doughty, 1979: 19). For some museums the loss of integrity came very soon after their establishment. Scarborough Literary and Philosophical Society (SLPS) provides a particularly dramatic example of the lifecycle of these institutions.

The gentry of Scarborough had been seriously discussing the establishment of a museum in the town since 1820. Little progress was made until this same group came together to form the Literary and Philosophical Society in 1827. Within two years they had built the remarkable Rotunda – a 'building in the round' designed specifically for the stratigraphic arrangement of fossils – an idea suggested by William Smith who had become increasingly attached to the town. Despite the free supply of stone from Sir John V.B. Johnstone, the Society's patron, the building, which was soon considered too small for meetings or the adequate display of the collections, placed a heavy burden on Society funds. It had been built with a loan of £500, secured by 19 signatures, on which interest was paid annually from subscriptions. Constantly short of money, the Society devised numerous ways to generate income but it was nearly a decade before the interior was finished. They were also unable to purchase important and desirable local fossils when they were discovered and so could neither fulfil their collecting obligation or raise their profile.

By 1842 the Society was approaching crisis. A circular was produced to rally support:

> the Council have to regret that the Institution does not receive that support from the town and neighbourhood generally which its acknowledged utility would seem to demand and further that owing to deaths, resignations and various other causes they seem to notice a considerable decrease in the amount of annual subscriptions.[3]

The crisis had deepened by 1848 when the initial loan on the building was recalled. The embarrassed state of the institution was then revealed in a public meeting chaired by Johnstone. The collections were found to be disorganized, poorly labelled and unattractively displayed, there was poor financial control and the membership was rapidly declining due to general dissatisfaction with the running of the institution.[4] While other society museums had increased their admission receipts since the opening of the railway, Scarborough's had declined since 1836. An attempted rejuvenation failed and by 1853 the Society saw merger with the more fashionable Archaeological Society as its only chance of survival. In its new form the Society and its museum survived into the twentieth century when management of the latter was passed to the town.

The neglect already apparent in Scarborough's collections was not uncommon even in 1850, particularly in the small museums now springing up. Edward Forbes, one of the new breed of professional geologists, speaking on the educational role of museums in 1854, described a fairly typical museum scene:

> Unfortunately not a few country museums are little better than raree-shows. They contain an incongruous accumulation of things curious or supposed to be curious, heaped together in disorderly piles, or neatly spread out with ingenious disregard of their relations. The only label attached to nine specimens out of ten is 'presented by Mr or Mrs So-and-so'; the object of the presentation having been either to cherish the glow of generous self-satisfaction in the bosom of the donor, or to get rid – under the semblance of doing a good action – of rubbish that had once been prized, but latterly stood in the way. (Rudler, 1877: 20)

Forbes' words were applicable to the private museums established before the philosophical revolution, and to those now established with public funds. They would be echoed throughout the remainder of the century and applied increasingly to the once objective collections of the philosophical institutions.

What the 'country museum', and the museum in Scarborough, lacked was the scientific expertise necessary to curate the collections. The Rotunda's first curator had been John Williamson, a local collector who remained in the post for more than 27 years. He passed on much of his collection to the new museum, but he was treated very much as an employee, and maintained on a paltry 'salary'; for him the museum was 'a labour of love'. By trade, Williamson was a market gardener and he had 'enjoyed no educational advantages' (Williamson, 1896: 6). Like others in the Society 'though not a philosopher in the higher sense he was a true worker'.[5]

Rudler (1877: 35) was one of many to emphasize the vital importance of staff to the success of a natural science museum. It was essential for the museum to have a post for a professional curator with 'an intelligent acquaintance with natural history' which could be maintained as successive individuals came and went. 'What a museum depends upon for its success and usefulness is not its building, not its cases, nor even its specimens, but its curator. He and his staff are the life and soul of the institution, upon whom its whole value depends' (Flower, 1889: 12). Flower[6] asserted the vital importance of staff at every museum opening; to him museums lacking a paid and knowledgeable curator were 'traps into which precious objects fall only to be destroyed'.

The role of John Phillips, in the Yorkshire Museum, is archetypal proof of this wisdom, but perhaps surprisingly it was not a view with which Phillips agreed. His experience had taught him that, in the presence of a paid keeper, the members soon showed 'a gentle acquiescence in growing indifferent to real study, and an upspringing demand for something more amusing, exciting or fashionable'; and the keeper, frustrated and overworked, gave his attention to other things (Orange, 1973: 40). Thus after he left the Yorkshire Museum Phillips repeatedly advised against the appointment of another paid keeper, despite the benefits he had derived from the post. Similarly in 1853 he advised the Scarborough Society to acquire a lowly paid caretaker who would keep the objects clean and that a 'really scientific man' could visit periodically.[7] The prominent geologist, Sir Roderick Murchison, was generally in agreement with Phillips' view. Though he had little experience of the functioning of museums, he hoped that the keeper would have sufficient knowledge to be able to point out the most remarkable fossils to strangers.[8]

Change of fashion was not something which could be fought against in the way Phillips suggested, but perhaps the Council of the Yorkshire Philosophical Society were aware of this as they soon appointed his replacement. They knew all too well that their success was in good measure due to Phillips who had 'very superior scientific attainments, is modest, sensible and popular, well contented with science and £100 a year'.[9] Inheriting an unparalleled knowledge of British geology from his travels with his uncle and guardian William Smith, he was also a gifted communicator (Phillips, 1844; Edmonds, 1982: 145). Pyrah (1988: 2) also believes that the relative success of geology in the nineteenth-century Yorkshire Museum was entirely due to the presence of a geologist in the post of keeper. 'When in the 20th century this tradition was broken growth of the geological collections virtually ceased . . . [they] came to be seen as an heritage from the past rather than as an actively evolving department'. Founded on the Kirkdale treasures,[10] rooted in local geology, and associated with Phillips until his death in 1874, it would have been difficult for the Society to shrug off its geological obligations even if it wanted to.

Consistency in staffing, and therefore commitment, was vital to the survival of geology in museums – without it even the larger institutions were liable to the problems of the smallest.

> It is . . . a dangerous thing for a public museum to depend thus upon the support or interest of a single individual, or even on a few amateurs, such as form our local natural history clubs: and it has indeed often happened that when the leading scientific spirit of a locality has been removed, the museum has degenerated, and lapsed into a state of neglect. (Rudler, 1877: 34)

All subsequent incarnations of the provincial natural history museum, whether privately or publicly funded, were subject to this same law (Flower, 1898: 55; Gill and Knell, 1988: 12; Torrens and Taylor, 1990: 197).

THE RISE OF MUSEUM GEOLOGY

The commentators of the late nineteenth and early twentieth centuries continued to repeat Forbes' concerns – neglect was widespread and particularly prevalent in geology collections. Museums were underfunded and misunderstood by their local community and government (Lankester, 1897: 21). The bottom of the syncline seems to have been reached in the 1860s (Manton, 1900; Scharff,[11] 1912).

By that time the once young and inspiring philosophers had become tired and conservative. The younger generation were not enticed into the old establishment but sought new and more fashionable enterprises. Natural history societies and field clubs boomed, in part spurred on by new opportunities for travel in the 1860s and 1870s (Anon, 1870a: 249; Allen, 1976: 164). Some of these societies also sought to establish museums but many had learnt of the problems of too many material encumbrances. However, unlike their Philosophical predecessors they did not need to develop their own museums. Instead they might follow the lead taken by Folkestone Natural History Society which, in league with the town council, renovated and took charge of an existing ailing museum (Anon, 1871: 381). Such societies continued to have an important role in the encouragement of local museums well into the twentieth century.

Whereas every philosophical society had had a museum, this was true of only 28 per cent of scientific societies in the 1880s and some of these were survivors from the earlier generation (Galton *et al.*, 1884). Ball *et al.* (1888: 123) found that half of existing museums had originated as society collections, and that half of these were now taken over by municipal authorities or trusts.

The current fashion for natural history included geology, and it was this that still dominated museum collections: in nearly 50 per cent of all museums, geology remained the largest collection (Figure 2.1) (Ball *et al.*, 1888: 114). There was quite even coverage of biology and geology (Figure 2.2), but the relative ease of

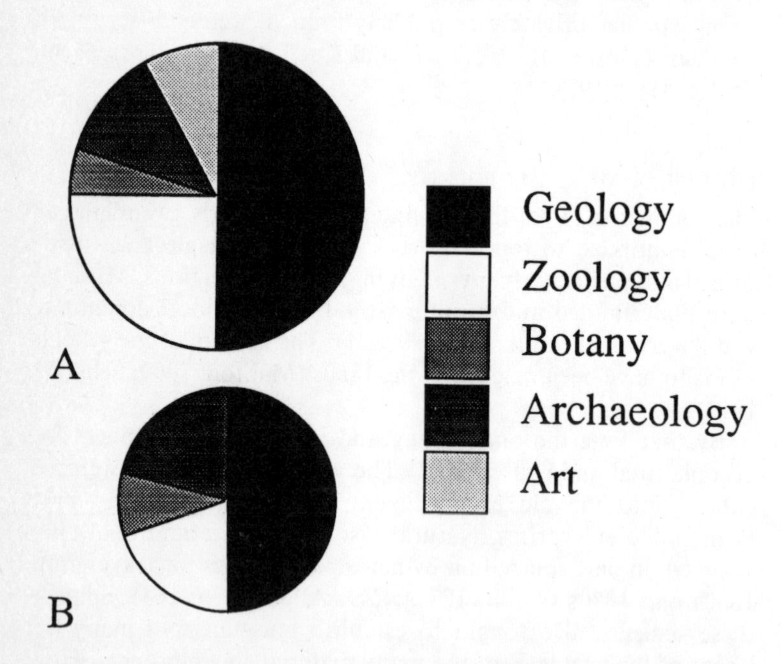

Figure 2.1: Largest collections in provincial museums in Britain in 1887 (Ball *et al.* 1888). A. General collections. B. Local collections (from a sample of 211 museums).

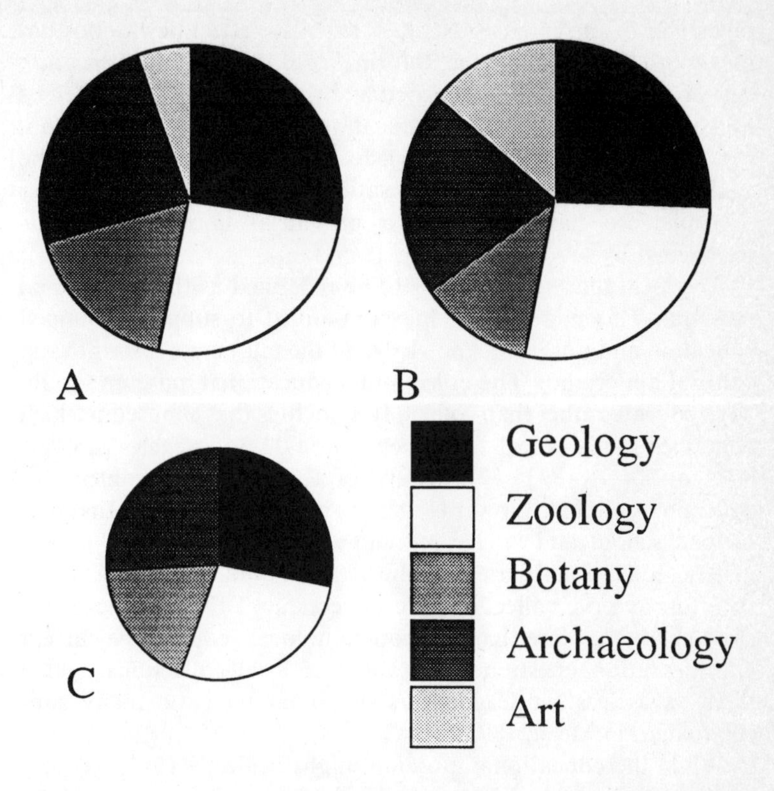

Figure 2.2: Representation of disciplines in museum collections in 1887 and 1914 based on the number of museums holding a collection in each discipline (Ball *et al.* 1887; Green *et al.* 1921). A: General collections 1887. B: General collections 1914 (from a sample of 134 museums). C: Local collections 1887. Note that areas cannot be summed as most museums held more than one discipline.

collecting and preserving geological material probably explains its dominance. The natural sciences continued as the major area of museum interest well into the next century (Green *et al.*, 1921: 269).

The natural history societies and their associated museums were seen as being of two types: those with higher scientific objectives and those rooted in formal and popular education (Gray, 1865; Anon., 1870b: 469; Flower, 1893: 37). Museum developments in the late nineteenth century were a reaction against the inadequacies of existing provincial museums, but also against the higher scientific objectives of the earlier societies. They also reflect new opportunities. With the passing of the Reform Act of 1867 and the Education Act of 1870 audiences broadened and the buzzword for museums for the next 70 years became education (Lankester, 1870; Jarman, 1963: 264; Bowen, 1981: 447–53). Joseph Hooker, speaking to the 1868 meeting of the British Association, stated that he could 'never remember to have heard of a provincial museum that was frequented by schools' (Hooker, 1869: lxiii).

The local museum was now to be redefined as the educational museum. This new type of museum aimed to supply a rounded education and not just a knowledge of the museum's scientific and cultural hinterland. 'The object of an educational museum should be to educate rather than collect. It is obvious that a museum which contained only local specimens would not teach geology' (Hutchinson,[12] 1893: 52). Aware of the practical problems of collection building, Flower (1893: 52) also recommended that municipal, school and village museums should only collect largely general educational series; 'nothing else should be attempted and therefore reserve collections are unnecessary'. Even at the Yorkshire Museum, more than any rooted in local geology, the current administration questioned the value of local collections, seeing them as useless for teaching as they contained too many gaps (Platnauer, in Meek, 1895).

While the educational museum might neglect its role as local archive others saw this as the provincial museum's priority (Green *et al.*, (1920: 271). These museums were to 'devote themselves to the thorough and complete working out of the productions of their own districts' (Ball *et al.*, 1888: 124). 'When a naturalist goes from one country to another his first inquiry is for local collections. He is anxious to see authentic and full cabinets of the productions of the region he is visiting' (Forbes (1854) quoted in Rudler, 1877).

> It seems to me also very proper to suggest that the great value of your museum is and ought to be in its departments which illustrate your own land and sea. General collections are not to be made or maintained except in places favourably situated.[13]

Only half of all provincial museums in the 1880s possessed specifically local collections (Ball *et al.*, 1888). General collections were seen as being important in the proposed National Museum of

Wales, but Rudler (1877: 19) felt that these must be kept within moderate limits, only finding a place in the museum through educational merit rather than simple aesthetics or rarity.

One of the most notoriously poor museums of the period existed in Canterbury (Anon, 1871: 381). It had been established in 1847, one of the first under the Museums Act of 1845, and had taken over the collections of the local philosophical society. Adopting a weapon popular with advocators of provincial museums, George Gulliver (1871: 35) of the town's East Kent Natural History Society launched a faintly disguised media attack on his local museum. 'The majority of them [provincial museums] throughout England present such examples of helpless misdirection and incapacity as could not be paralleled elsewhere in Europe'.

The Canterbury Society believed they possessed the knowledge required to run a natural history museum and were supporters of Folkestone's approach. But they were intent on pursuing an entirely different objective to that on which the collection was originally founded – now the goal was public education. 'Nobody in his senses can suppose that it is either desirable or practicable for a provincial society to attempt an imitation of the vast and boundless metropolitan institution'. Thus if material in Canterbury had survived the decline of the local philosophical society, and the neglect of the town council, would it also survive the reorganization imposed by the newly fashioned natural history society? In rescuing the useful series, a good deal of 'rubbish' was likely to be encountered. The solution to this problem: 'sell it if you can, or give it away; but by all means get rid of it, and that swiftly; to which end a bonfire might be the best thing' (Gulliver, 1871: 36). While local collections and general educational series might survive, important material brought back from further afield might be considered useless – a narrow-minded view of existing collections which has persisted in the increasingly locally focused museums of the twentieth century (Knell and Taylor, 1991). Old collections might survive but each re-use brought with it a cleansing which was as destructive as it was revitalizing.

Scharff (1912: 12) saw the Great Exhibition of 1851 and the development of the government's Department of Science and Art, under Sir Henry Cole, as providing the basis for the widespread adoption of educational museums. After long debate it was through this Department that the Natural History collections of the British Museum at last found freedom and opportunity for growth in South Kensington (Flower, 1889: 10). With it came a renaissance of interest in palaeontology linked to growing interest in Darwinism (Woodward, 1900: 37). The Department also encouraged the formation of classes in a large number of sciences; 'it introduced large numbers of citizens to the marvels and potentialities of science' (Boswell, 1941: xxxviii), and seemed capable of advancing those provincial museums established since the Museums Act 1845. Scharff's (1912: 12) view was common:

> The old popular conception of a museum as a repository for curiosities has passed away, and a new order of things has been established. Whereas not long ago museums still existed, containing nothing more than an ill-assorted mass of rubbish . . . but almost every museum started its early career in that manner.

However, his words express an optimism about museums that fails to recognize past lessons and which was to be undermined by the difficult decades ahead.

While the natural history societies and field clubs were slightly more egalitarian than their predecessors, the middle classes still remained in control. Flower (1889: 14) was keen to see museums reach out to those parts of the community who through education and opportunity lacked the chance of profound study but might be more generally engaged by museum objects. Such a view was not however universal. Henry Woodward, President of the Museums Association and stalwart of the Natural History Museum for 43 years, felt 'that the "man in the street" did not at present seem to be a very hopeful subject in London. He came into museums chiefly for warmth and shelter, and usually brought a good deal of dirt in with him' (Manton, 1900: 75). This snobbery was still quite deep rooted in the Natural History Museum, and even revealed in Bather,[14] but appalled the more enlightened curators of the provinces (Manton, 1900: 76).

In the early years of the Museums Association (in the 1890s) natural scientists dominated proceedings. At its annual meetings geology was a popular subject for focused discussion and for the illustration of more general principles. What emanated from the meetings was widely adopted as good practice. An increasingly educational role for museums brought with it much discussion of the most effective methods of geological display. In the decades when geology was fashionable and fairly exclusive there was no apparent need to communicate effectively as museum displays illustrated current news items and a curator or member was always on hand to inform the visitor. Now the museum audience was changing and natural history display needed to be organized so as to meet the needs of its two visitor groups so that 'students would be uninterrupted by the ignorant curiosity of the ruder class of general visitors' (Gray, 1865: 77). The old approach had been to provide a vocabulary:

> There are some, however, who imagine that nothing is to be learned from a museum except a catalogue of names; but this, even were the statement true, is surely an unreasonable complaint. If the volume of nature is worthy of being read, its vocabulary must deserve to be studied. No one can learn the names of these objects, without first acquiring, at the same time, some knowledge of their properties; and no one can discuss the properties without possessing some knowledge of the names. (Yorkshire Philosophical Society, 1828: 14)

Geologists seemed to be searching for the *one* effective and appropriate way to display collections. Consequently, exhibits were conservative and inflexible. Ball *et al.* (1888) showed that provincial museums almost universally adopted the same methods of display. For example, fossils, if they were organized at all, were arranged stratigraphically and then zoologically within the stratigraphic group – a method of display devised by William Smith a century earlier. Organization was generally systematic, geographical or temporal (stratigraphic) (Rudler, 1877; Dawkins, 1890: 38; 1892: 22). To Woodward (1900: 33) there was no difference in the requirements of arrangement of objects in the galleries as compared to those in the store. Even some of those working to break from systematic arrangement still saw it as being important in more academic museums (Bather, 1924: 133), though Bather (1896: 94) had earlier admitted that in provincial museums these were often based on outmoded classifications which more reflected the date of foundation than contemporary thinking.

John Edward Gray (1865: 76) having dedicated his life to the creation of order out of chaos in the zoology collections of the British Museum, realized at the age of 65 that his work had benefited the scientist at the expense of the public. While systematic arrangement was to stay, he called for an end to those comprehensive display series which placed similar species together only to create confusion and dismay.

In the philosophical age the geology collection provided a resource for the illustration of lectures, which were also supported by coloured maps, drawings, sections and diagrams. But these illustrative materials were little used in museum displays. Hutchinson (1893: 52) was one curator beginning to break the old mould by promoting the liberal use of pictorial illustration, models, casts, descriptive labels and the availability of reference texts in educational museums. But still he arranged the collections in the limited number of ways then seen as appropriate. In addition there remained the question of how the local collection should be used. 'A museum which remains entirely local misses something of high educative importance' (Green *et al.* (1920: 271). To meet these educational objectives museums needed to place specimens in their wider context. 'A larger museum might prefer to have a fairly representative collection of geological and zoological specimens ... this should be kept altogether separate from the Local Museum, and must, of course, be arranged in strictly systematic order' (Meek, 1895: 156). A universal rule had been adopted separating local and general collections. The latter were divided into 'type' or 'typical' series relating to the major fossil, mineral and rock groups. The approach had been developed in Richard Owen's concept of the 'index museum' which he planned to incorporate into the new Natural History Museum in South Kensington (Stearn, 1981: 57).

A fundamental and influential shift in thinking was taking place at the Natural History Museum at this time. Buoyed up by renewed interest in evolution in the wake of Owen's retirement, the Museum's scientists began to consider how the barrier which divided its research in two could be dismantled. 'Perpetuation of the unfortunate separation of palaeontology from biology, which is so clearly a survival of an ancient condition of scientific culture, and for the maintenance in its integrity of the heterogeneous compound of sciences which we now call "geology" the faulty organization of our museums is in a great measure responsible' (Flower, 1889: 11). Attempts to break with tradition were made as recent species and genera were introduced into the systematic arrangement of fossil cases. The first was when Lankester added fossil genera to Flower's Cetacean gallery, a gallery which today, despite complete revision, continues to maintain this link between the two sides of the

Museum (Woodward, 1900: 42). In his important, though largely overlooked, essay, Rudler (1877: 30) also suggests the potential benefits of combining the display of recent and fossil species.

A more advanced approach to geology display was promoted by Bather in the 1920s. There were still some who felt 'there was nothing duller in the world than the specimens of fossils of various kinds to be seen in the ordinary museums' (Ridley in Bather, 1924: 140). For these non-believers Bather described how the intellectual content of fossils could be unlocked to reveal the dynamics of former lives – adaptation, predation, defence, disease, death and so on. To assist those new to the subject Bather recommended contemporary science texts which might provide models for illustration in display. The newly inspired Ridley expressed concern at getting an adequate range of material, to which Bather remarked, perhaps shortly, that Ridley's local Red Crag would provide all any museum might need; 'the production of a series of fossils did not depend so much on the collection as on its curator'. Smith (1897: 65) earlier expressed the benefits of following the logical organization of a textbook and one that would meet the needs of local schools. Bather (1896) in 'How can museums best retard the advance of science?' suggested that museums should not be given over to serried rows of specimens but to ideas, such as Darwinian evolution, 'which make people think'.

THE FALL OF MUSEUM GEOLOGY

From the 1920s museum geology was once again beginning to slump. The reorganization of the Science and Art Department prior to the Great War was widely blamed for thrusting the science once more into a period of general decline (Boswell, 1941).

During the first half of this century, numerous writers were attempting to explain and reverse this decline (Bather, 1924; North, 1942; North, *et al.*, 1941); forty years later another generation of museum geologists were explaining the decline in similar ways and suggesting surprisingly similar solutions (Doughty, 1981; Brunton *et al.* 1985; Knell and Taylor, 1989).

The loss of material from the 1920s onwards, much of it dating back to the earliest days of local geological exploration, was remarkable. At this time there was no safety net: museums had not been sufficiently professionalized to be concerned about disposal;

curators were increasingly less informed about geology; and the watchdog Geological Curators Group was not to appear on the scene for fifty years. In south eastern England, Raymond Casey, a palaeontologist then working for the Institute of Geological Sciences (formerly and subsequently, the Geological Survey), witnessed this destruction. He rescued what he could and redistributed it to universities and schools. This included the whole of Tunbridge Wells Museum's collection – which was twenty years later found intact in the Geological Museum (Gill and Knell, 1987). Neglect and loss through sale, dumping, burial and theft was regrettably commonplace (Knell, 1987: 7).

> Unsatisfactory geology collections are more common than they should be, and they are to be found in museums where the local conditions are such that geology should be receiving special attention . . . .But if neglect, pure and simple, has to be given as one of the reasons why museums have failed in their duty towards geology, careful preparation and scientific arrangement have not always been followed by the results that curators may have anticipated. (North, 1942: 249)

North[15] (1942) believed that geological specimens were no longer understood.

> As such they are handled roughly and left lying about until what little individuality they may have had in colour or in texture is obscured by dust, so what ultimately passes for a 'geological collection' in some museums is an agglomeration of dirty, inadequately labelled, and ill-chosen specimens, that might as well have been shovelled from a pile of builder's rubbish or from a heap of roadstone for all the interest they are likely to create.

Allan[16] (1942, 58) suggested that the inherent durability of geological materials worked against the interests of these collections.

> Requiring less attention, they got it, and, while the passage of time saw new material and new methods of preparation and exhibition introduced into other natural history departments, the rocks and the fossils remained intact and inert, sometimes almost invisible beneath the gently accumulating layers of dust.

Modern geologists complain of the poor media coverage the subject achieves. Boswell (1941: xxxvi) too felt that low public interest in geology could in part be blamed on the media preoccupation with frivolous and disastrous geological stories rather than disseminating information about scientific advance. However, it would be wrong to suggest as Allan (1941: 60) did, that geological galleries of the past thrived because society was deprived of the cheap entertainments of the modern day, ignoring as it does massive social changes and cheap entertainments of the past!

While admitting that the mid-Victorian interest in field clubs demonstrated that it was 'still possible to find close at hand numerous problems awaiting solution', there was also a feeling amongst some geologists that there was then less to be achieved locally (Boswell, 1941: xxxvi).

Despite the efforts of Bather and others, North felt that geology in museums was still failing to communicate effectively with the public. The problem now was that geology had no obvious connection with everyday life or common knowledge; it had lost its sense of novelty and wonderment; and it appeared to have no utility. Systematic displays were still prevalent in museums; 'they have a system but no soul; the specimens have names but are devoid of meaning' (North, 1942: 249).

Following Bather's approach, North set out to reawaken interest in geology through museum display. He attempted to educate curators in the hope that they would educate the masses. In 'Why geology?' he tried to relate geology to other areas of museum interest such as art and archaeology. North believed that geology in museums could no longer 'rely upon a novelty that is worn out'; it needed to tell its own story, and, like those who preceded him, he believed that the provincial museum should be teaching general principles. Local material was an appendix to this general story which should be used to give an overview of local geology. In 1928 he described the indispensability of a case introducing the science of geology to the museum visitor. North used supporting materials to illustrate and explain, much as Hutchinson and Bather had done before him. However, despite North's undoubted eloquence, his interpretation was seen as being rather dry and technical (see Bather in North, 1928b: 16). North's ideas influenced a generation of curators and his suggested improvements to local geological exhibits found their way into many British museums where they largely remained until the 1970s (North *et al.*, 1941). Geological interpretation was sober and educational; attempts to enliven it were not always welcomed: 'Popularity is sought by catchpenny methods, such as the gem-stone exhibit, the fluorescent lamp and the diorama peepshow – mirages to lure the wandering visitor into a desert of desk cases' (Allan, 1942: 60). The teaching of general principles made geology displays throughout Britain largely identical. To a more travelled modern audience this would only result in tedium.

TO RISE OR TO FALL

Major periods of growth and decline have been suggested here, but these are simply national trends which have been prevalent at certain times; they in turn reflect wider changes in society (democratization, the economy, fashion, education). These were superimposed on more local factors: dominant personalities, urban growth and prosperity, interactions between neighbouring towns, and so on. Geological interest mushroomed in those towns which had acquired sufficient size, or persons of influence, to spur activity; the concept of critical mass is useful here. But as Rudler pointed out, this interest was likely to be ephemeral. Nor should it be believed that a local growth of interest would be good for the local geological heritage. The example of Canterbury Museum suggests that each new blossoming of interest brings with it a cleansing of old collections and a further weakening of links with the 'Heroic Age'.

The new era for museum geology in Britain dawned with the formation of the Geological Curators Group in 1974, which commissioned Phillip Doughty to report on the status of geology collections in the UK (Doughty, 1981). Museums were now more professional, better funded and staffed, and more attuned to the needs of heritage conservation. As such, it is not surprising that Doughty's findings caused widespread shock and disbelief.

Curators were already aware of disarray in their own museum and perhaps a few others but no one had the overview. A veneer of professionalism covered the warts of a museum system incapable of carrying through its mission; in a culture built on local pride, failures were seldom discussed. But 'the profession', as it is now known, is a very recent invention and was predated by more than a hundred and fifty years of poorly resourced amateur (i.e. without training, method or standard) involvement. Museums were powerless to resist fashion, the economic realities of museum provision or changing notions of what a museum should be. Disarray accompanied the growth of collections.

While individual collections may have found order for a few years, most have probably spent much of their time in total or partial chaos, or simply in an unmaintained state. Many of the neglected collections examined in the 1980s were wrapped in 1920s newspaper; often this was the paper in which they were wrapped when donated. It is not difficult to visualize the causes of neglect and loss, because they still exist and threaten collections – a further application of the geologist's Principle of Uniformitarianism.

In the 1990s geology in museums is more politicized than in the past. It is more capable of protecting itself, although it can never be immune from those factors which create loss and decline. A key element in the success of geology in museums has been the provision of the specialist curator; but the museum geologist is as threatened as his or her collection. As museums staff themselves meet current or perceived obligations there is a danger that the museum specialist may suffer further decline (Knell, 1995). There is sufficient evidence to show that where there is good provision in terms of staff, museum geology has public appeal which few disciplines can match. Geologists in the 1990s have also discovered that North was in error in believing that the 'wonderment' which once surrounded geology cannot be recovered. The facts revealed in the 'Heroic Age' caused considerable public interest. Those facts remain to be discovered by every generation – usually as children – for whom they are just as fascinating. Fortunately, modern geologists have been released from the belief that displays must 'teach' principles; they know more about how visitors learn in the museum but also that their most important role is to inspire. Geology exhibits are also beginning to take advantage of the technology which can enhance its potential to communicate imaginatively. At the same time, provincial museums have rediscovered local geology and thrust this to the fore; in doing so many have rediscovered their roots.

The current renaissance in museum geology is in large part due to a renewed sense of purpose. There has been little growth in provision. The 'Catch 22' for geology in museums is that without provision (i.e. a museum geologist) there is no demand; if there is not demand there will be no provision. There is no reason to believe that the current renaissance will be any less ephemeral that those which preceded it.

ACKNOWLEDGEMENTS

I would like to thank the following for their help with archival materials: Linda Maggs, Yorkshire Philosophical Society; Stella Brecknell, Oxford University Museum; Ros Palmer, Scarborororough Museums; and the librarian at Scarborough Public Library. Thanks also to Hugh Torrens for telling me about the Perceval.

NOTES

1 F.W. Rudler, Professor of Natural Science in the University College of Wales previously and subsequently Assistant Curator and Curator, respectively, at the Museum of Practical Geology, London.
2 Letter from John Dunn, Secretary of Scarborough Literary and Philosophical Society to John Phillips, Keeper of the Yorkshire Philosophical Society, 1 March 1830. Oxford University Museum, Phillips Archive 1830/6.
3 Circular drawn up by the J.R. Hulme, Secretary of Scarborough Literary and Philosophical Society, 10 January 1842. In SLPS minute books, Scarborough Public Library.
4 SLPS Minute book, meeting of 5 October 1848, Scarborough Public Library.
5 W.C. Williamson in a letter John Phillips dated 7 October 1868, remarking on the death of William Bean, Oxford University Museum, Phillips Archive 1868/73.
6 W.H. Flower, Director of the Natural History Museum, London.
7 Letter from J.V.B. Johnstone to J. Leckenby, 19 March 1853 (427.39.1); Letter from J. Phillips to J.V.B. Johnstone, 22 March 1853 (183.58), in Scarborough Rotunda Museum.
8 Letter from J.V.B. Johnstone to J. Leckenby, 24 March 1853, in Scarborough Rotunda Museum, 427.39.2.
9 Letter from W.V. Vernon Harcourt to Lord Milton, 18 January 1831. Published in Morrell and Thackray (1984).
10 See W. Buckland (1822) 'An account of an assemblage of fossil teeth and bones discovered in a cave at Kirkdale', *Phil. Trans. R.S. Lond.*, 122: 171–236.
11 R.F. Scharff, Keeper of Natural History, National Museum, Dublin.
12 J. Hutchinson, founder of Haslemere Educational Museum, Surrey.
13 Letter from J. Phillips to J.V.B. Johnstone, 22 March 1853 (183.58) in Scarborough Rotunda Museum.
14 F.A. Bather, Assistant Keeper, then Keeper of Geology at the Natural History Museum, London.
15 F.J. North, Keeper of Geology, National Museum of Wales, Cardiff.
16 D.A. Allan, Director of the Free Public Museums, Liverpool.

BIBLIOGRAPHY

Allan, D.A. (1942) 'President's address for 1941: geology in museums', *Proceedings of the Liverpool Geological Society*, 18: 57–69.

Allen, D.E. (1976) *The Naturalist in Britain: A Social History* (London: Penguin).

Anon. (1870a) 'Natural history in schools', *Nature*, 2: 249–50.

Anon. (1870b) 'Natural history societies', *Nature*, 2: 469–70.

Anon. (1871) 'Natural history museums', *Nature*, 3: 381–82.

Ball, V. *et al.* (1888) 'Report of the committee on provincial museums', *Report of the British Association for the Advancement of Science for 1887*, pp. 97–130.

Bather, F.A. (1896) 'How may museums best retard the advancement of science?', *Report of the Proceedings of the Seventh Annual General Meeting of the Museums Association*, pp. 92–105.

Bather, F.A. (1908a) 'Visit to the palaeontological exhibit in the science hall, Franco-British Exhibition', *Proc. Geologists Association*, 20: 539–47.

Bather, F.A. (1908b) 'The preparation and preservation of fossils', *Museums Journal*, September: 76–90.

Bather, F.A. (1924) 'Fossils as museum exhibits', *Museums Journal*, 24: 132–41.

Boswell, P.G.H. (1941) 'Anniversary address of the president: part 1: The status of geology: a review of present conditions', *Quarterly J. Geological Society*, 97: xxxvi–lv.

Bowen, J. (1981) *A History of Western Education: 3. The Modern West Europe and the New World* (London: Methuen).

Brunton, C.H.C., Besterman, T.P., and Cooper, J.A. (1985) *Guidelines for the Curation of Geological Materials*, Geological Society Miscellaneous Paper no. 17 (London).

Buckland, W. (1836) *Geology and Mineralogy considered with reference to Natural Theology*, vol. 1 (London: Pickering).

Dawkins, W. Boyd (1890) 'On museum organisation and arrangement', *Report of the Proceedings of the First Annual General Meeting of the Museums Association*, pp. 38–45.

Dawkins, W. Boyd (1892) 'The museum question', *Report of the Proceedings of the Third Annual General Meeting of the Museums Association*, pp. 13–24.

Desmond, A. (1989) *The Politics of Evolution* (Chicago: University of Chicago Press).

Doughty, P.S. (1979) 'The state and status of geology in United Kingdom museums', in M.G. Bassett, (ed.) *Curation of Palaeontological Collections*, Palaeontological Association Special Papers in Palaeontology no. 22 (London), pp. 16–26.

Doughty, P.S. (1981) *The State and Status of Geology in United Kingdom Museums* (London: Geological Society).

Edmonds, J.M. (1982) 'The first "apprenticed" geologist', *Wiltshire Archaeological and Natural History Magazine*, 76: 141–54.

Flower, W.H. (1889) 'Museum organisation: presidential address to the British Association for the Advancement of Science' in W.H. Flower, (1898), pp. 1–29.
Flower, W.H. (1893) 'Modern museums: address to the annual meeting of the Museums Association' in W.H. Flower, (1898), pp. 30–53.
Flower, W.H. (1898) *Essays on Museums* (London: Macmillan).
Galton, F. *et al.* (1884) 'Report of the local scientific committee', *Report of the British Association for the Advancement of Science for 1883*, pp. 318–45.
Gill, M.A.V., and Knell, S.J. (1988) 'Tunbridge Wells Museum: geology and George Abbott (1844–1925)', *Geological Curator* 5(1): 3–16.
Gray, J.E. (1865) 'Address of the president of the botany and zoology section', *Report of the British Association for the Advancement of Science for 1864*, pp. 75–86.
Green, J.A. *et al.* (1921) 'Museums in relation to education – final report of the committee', *Report of the British Association for the Advancement of Science for 1920*, pp. 267–80.
Gulliver, G. (1871) 'On the objects and management of provincial museums', *Nature*, 5: 35–6.
Hooker, J. (1869) 'Address of the president', *Report of the British Association for the Advancement of Science for 1868*, pp. lviii–lxxv.
Harrison, J.F.C. (1971) *Early Victorian Britain, 1831–51* (London: Fontana).
Hutchinson, J. (1893) 'On educational museums', *Report of the Proceedings of the Fourth Annual General Meeting of the Museums Association*, pp. 49–63.
Jarman, T.L. (1963) *Landmarks in the History of Education* (London: John Murray).
Jones, R.E. (1980) 'Evolution, creationism and science education', in K. Allen and D. Briggs (eds) *Evolution and the Fossil Record* (London: Belhaven).
Knell, S.J. (1987) 'Geology collections in the south east', unpublished report of the Area Museums Council for South Eastern England.
Knell, S.J. (1995) 'Collection managers: the final insult?' *Geological Curator,* 6(3): 125–7.
Knell, S.J., and Taylor, M.A. (1989) *Geology and the Local Museum* (London: HMSO).
Knell, S.J., and Taylor, M.A. (1991) 'Museums on the rocks', *Museums Journal*, 91(1): 23–25.
Lankester, E.R. (1870) 'Science at school boards', *Nature,* 3: 161–2.
Lankester, E.R. (1897) 'Presidential address', *Report of the Proceedings of the Eighth Annual General Meeting of the Museums Association,* pp. 19–29.
MacLeod, R.M. (1983) 'Whigs and savants: reflections on the reform movment in the Royal Society 1830–48' in I. Inkster and J. Morrell, (eds) *Metropolis and Province: Science in British Culture, 1780–1850* (London: Hutchinson), pp. 55–90.

Manton, J.A. (1900) 'A rambling dissertation on museums by a museum rambler', *Report of the Proceedings of the Eleventh Annual General Meeting of the Museums Association*, pp. 65–80.

Meek, A. (1895) 'The development of a local museum', *Report of the Proceedings of the Sixth Annual General Meeting of the Museums Association*, pp. 149–61.

Morrell, J. (1994) 'Perceptual excitement: the Heroic Age of British geology', *Geological Curator*, 5(8): 311–17.

Morrell, J. and Thackray, A. (1981) *Gentlemen of Science: Early Years of the British Association for the Advancement of Science* (Oxford: Clarendon).

Morrell, J. and Thackray, A. (1984) *Gentlemen of Science: Early Correspondence of the British Association for the Advancement of Science*, Royal Historical Society, Camden 4th series (London).

North, F.J. (1928a) 'Geology and the museum visitor', *Museums Journal*, 27: 271–6.

North, F.J. (1928b) 'Geology in relation to the small museum', *Museums Journal*, 31: 8–17.

North, F.J. (1942) 'Why geology?', *Museums Journal*, 41(11): 249–56.

North, F.J., Davidson, C.F., and Swinton, W.E. (1941) *Geology in the Museum* (Oxford: Museums Association/Oxford UP).

Orange, A.D. (1973) *Philosophers and Provincials: The Yorkshire Philosophical Society from 1822 to 1844* (York: Yorkshire Philosophical Society).

Perceval, S.G. (1871) 'The Museums of the Country', *Nature*, 4: 367.

Phillips, J. (1829) *Illustrations of the Geology of Yorkshire: part 1 The Yorkshire Coast* (London).

Phillips, J. (1844) *Memoirs of William Smith* (London: John Murray).

Pyrah, B.J. (1988) *The History of the Yorkshire Museum and its Geology Collections* (York: Sessions).

Rudler, F.W. (1877) 'On natural history museums', *Y Cymmrodor*, 1: 17–36.

Rudwick, M.J.S. (1985) *The Great Devonian Controversy* (Chicago: University of Chicago Press).

Scharff, R.F. (1912) *The Aims and Scope of a Provincial Museum* (Belfast: Library and Technical Construction Committee).

Smith, H.J. (1897) 'Popular museum exhibits', *Report of the Proceedings of the Eighth Annual General Meeting of the Museums Association*, pp. 63–8.

Stearn, W.T. (1981) *The Natural History Museum at South Kensington* (London: Heinemann).

Torrens, H.S., and Taylor, M.A. (1990) 'Geological collectors and museums in Cheltenham 1810–1988: a case history and its lessons', *Geological Curator*, 5(5): 175–213.

Whitby L and PS (Literary and Philosophical Society) (1826) *Third Annual Report*.

Williamson, W.C. (1896) *Reminiscences of a Yorkshire Naturalist* (London: Redway).

Woodward, H. (1900) 'Presidential address', *Report of the Proceedings of the Eleventh Annual General Meeting of the Museums Association*, pp. 25–44.
Yorkshire Philosophical Society (1828) *Annual Report of the Yorkshire Philosophical Society for 1827*, p. 14.

3

Presenting Science as Product or as Process: Museums and the Making of Science

KEN ARNOLD

INTRODUCTION

Set against a starry backdrop on the cover of the Science Museum's 'Catalogue Collection Christmas 1994' is a picture of The Orrery – 'an ingenious clockwork construction kit . . . accompanied by a 15 minute cassette by world famous astronomer Patrick Moore'. It costs £9.99. Next to it is another of The Original Orrery – 'A mechanical model of the solar system in the Science Museum collection'. This one is 'Not for sale!'. More than just a marketing gimmick, the pairing stands as an emblem for the tendency in museums to present science as a series of products rather than processes – to narrow down the 'open-ended activities of science into objects and knowledge' (Porter, 1993: 25).

The shop, of course, is not the museum; but this image makes crystal clear just how common and easy it is for institutions formed around collections of objects to focus on commodities. The fact that this is a sales rather than exhibition catalogue is, however, entirely relevant, since, as Alan Morton has pointed out, the increasing dominance within museums of commercial contexts means that visits have themselves become a type of commodity (Morton, 1992: 137). As a result, science in museums often appears simply as one commodity within another, in this case a commodity which is the direct product of what scientists know. But as Steven Shapin has pointed out, what this scheme leaves out are all the questions of 'how, with what confidence, and on what bases, scientists come to know what they do' – that is, what science is actually like in the making (Shapin, 1992: 28).

The theme of this paper is the tension within science museums between presenting the subject in terms of its products and its processes. In the first part, I will argue that for much of their history museums have played active roles in the creation and dissemination of scientific ideas; but that as the relationship between science and museums changed in the nineteenth century, the latter came to figure more as graveyards of scientific history, or less morbidly, as its trophy chests. And in much the same way that natural history and art museums had for much longer displayed their treasures, so science museums came simply to put out for show their finest artifacts.

In the last decade or so, however, and for a variety of reasons both internal and external to the curatorial profession, many museums have attempted to move away from this model. In the second part of the paper I examine how a number of exhibitions have moved the central question on from what science does to how scientists do it. Here I will look at a series of projects that have uncovered just how central arguments over ideas are to the process of scientific enquiry, how science has to be broadcast as well as formulated, how, for better and worse, it is involved in the lives of ordinary people, and how seemingly immutable ideas commonly change over time. Together, these exhibitions suggest that science is much better understood as an accumulation of a range of philosophical, cultural and social processes.

MUSEUMS AND THE PRODUCTION OF SCIENCE

From the very commencement of their early modern history, museums have formed one of the privileged sites of scientific enterprise. Thus in the case of their birth, or rather rebirth, in sixteenth and early-seventeenth century Italy Paula Findlen has recently described how museums played host to the transition of old Aristotelian science into new experiment-based enquiries. What she has shown is how collectors like Ulisse Aldrovandi and later Johann Kircher used their collections and the cabinets and studies that housed them as tools by which to capture and introduce nature into a controlled environment where it could be experienced and experimented upon. Caught up in the civilizing process that Italian society was then experiencing, gathering collections and making scientific use of them constituted a form of behaviour that gave those who pursued it a certain social distinction. Founded on curiosity and civility, then, these early museums were established as theatres of scientific activity (Findlen, 1994).

Taking as a second example the nature of English museums in the seventeenth century, evidence for their centrality in the creation of scientific knowledge is again abundant. Contemporary correspondence, philosophical journals, accounts of learned lectures and even museum catalogues all clearly indicate that far from simply being repositories for objects, early English cabinets of curiosities were also used as studies, laboratories, demonstration halls and lecture theatres. As well as being carefully observed, the treasured rarities gathered there were also not uncommonly put in water, magnetized, set fire to and tasted. And the knowledge that was teased out of these collections of curiosities was used in a whole series of research programmes that came together within the walls of both private and institutional cabinets. These museums were put to work in creating natural historical surveys of British counties, in gathering information about the material cultures of countries discovered during voyages of exploration, in investigating the effectiveness of materia medica, and in inventing philosophical languages based on solid material facts (Arnold, *et al.*, 1993; and forthcoming).

Much later in the nineteenth century, though unrecognizably altered from their origins in Renaissance cabinets, museums were still, as Sophie Forgan has put it, 'at the forefront of knowledge' – 'a characteristic *locus* of scholarship and research' (Forgan, 1994: 140). John Pickstone has in fact gone so far as to characterize a whole era of scientific enquiry (roughly from the end of the seventeenth century to the beginning of the nineteenth) as 'museological' – a style of analytical understanding 'historically embedded in the emergence of specialist professional groups with command over large bodies of material' (Pickstone, 1994: 113).

It was during this period that the taxonomic sciences of zoology, botany, mineralogy, nosology, dermatology and even phrenology emerged, bringing with them disciplinary boundaries defined by ever larger and more lavish multi-volume publications. And the professional skills required to organize all the information and samples they were based on were none other than those of the curator. According to Pickstone, the museological practices they employed – ones which even characterized subjects like chemistry, primarily pursued far away from the museum proper – came to a head in Paris between 1793 and 1795, when a whole series of institutions and museums devoted to this methodology were set up in the name of the state and its citizens. The details of this cultural moment are less important than Pickstone's main contention, namely that museums were central to the active creation of scientific knowledge.

Not only crucial to its creation, museums also played a large part in scientific education. Thus the Conservatoire National des Arts et Métiers, set up in Paris in 1794, was used as 'a place of demonstration where craftsmen came to exhibit whatever bore the "stamp of usefulness" ' (Ferriot, 1992: 79). Indeed, scientific instruction had, since the seventeenth century, increasingly been based on lectures which drew on museum collections for their illustrative and demonstrative material. Thus the art of the eighteenth-century travelling lecturers with their cases of instruments and show pieces became institutionalized in colleges and academies.

During the course of the nineteenth century this model of science in which its production was intimately linked with museums crumbled and all but disappeared. Having occupied a spot near the centre of the scientific stage, museums were primarily displaced by laboratories. It was the latter that came to define an experimental ideology of command, control and manipulation, which consequently left in the shadows the knowledge based on classification that had been produced by studying museum collections (Pickstone, 1994: 131–2). The Parisian Conservatoire National was typical of a number of institutions that subsequently lost their importance, witnessing its museum-workshop progressively turn into the dead weight of a 'collection'. And a formerly important institution such as the Jermyn Street Geological Museum declined so much that by 1913 it was declared London's loneliest site – a fine place 'for growing beards' (Forgan, 1994: 156). Having been unchallenged as centres of scientific knowledge and education, museums began to appear more like scientific mausoleums.

The science that had been actively practised in museums gradually came to be frozen in place as a series of landmarks signalling a progressive history of science, the modern offspring of which had since taken up residence in other types of institution. Less and less places where science was shaped or argued over, museums appeared instead as storehouses for monuments to past triumphs. A considerable amount of celebration of contemporary scientific achievement was still presented to the general public, especially towards the end of the nineteenth century; but this was done in 'modern' industrial expositions rather than old-fashioned museums. The former were, as Joel Bloom has described them, the 'theme parks of their day' – where 'steam engines and other mechanical marvels [were paraded as . . .] the symbols and the reality of progress' (Bloom, 1992: 15). The philosophical undergirding for this view of science came in the form of logical positivism in which the success of new science was judged primarily in terms of how much old science it made obsolete. It was this form of progress that increasingly ensured that the only fit place for old science was a museum, and by the end of the nineteenth century that term had taken on a decidedly pejorative meaning.

The legacy of this positivistic philosophy has lasted late into the twentieth century, and is still embodied in the standard museum presentation of much science and technology, and possibly particularly transportation. Commonly, the only story told is that of progress, with displays habitually following a 'linear pattern of development, from earliest to latest and from inventor to market (less often onward to actual use)' (Porter, 1993: 25). Exhibitions of scientific objects, even when shown working, have tended not to be related to the environments in which they are found, thereby suppressing their 'functional or historical context and [isolating them] from economic, social and technological factors' (Parkyn, 1993: 31). Divorced from these social realities, visitors to exhibitions have no means of understanding how scientific and technological knowledge is actually produced. Instead, most museums touching on the history of science – those, for example, in Oxford, Cambridge, Florence, Utrecht and Leyden – have tended to embody an antiquarian interest in the subject, showing off their objects as if they were a fine collection of fossils.

One does not have to go far into almost any museum of science and technology in the country to find examples of this type of philosophy. Thus the magnificent East Hall in London's Science Museum parades rows of awe-inspiring steam engines. But presented in the fashion they are, as gleaming products of a bygone era, they do little to suggest the revolutionary impact that their power had first on this country and eventually on the entire world. Even those that are seen gently flexing their muscles do so to no visible effect, so that it is hard to connect them with work of any kind, let alone the operation of a mine or factory production line.

On the second floor in the chemistry gallery, to take one more example, visitors are presented with case after case of none-too-edifying flasks, furnaces, retorts and weighing machines. Even traces of more recent chemical investigations – working models of a smoking machine and a chromatograph – are presented against neutral backgrounds accompanied only by far-from-informative labels dwelling mainly on technical specifications. To be fair, this gallery does include reconstructions of a government laboratory *c.* 1897 and another from 1965; but even here it is difficult to find out, or indeed imagine, who might have been doing what in them, to what end and for the sake of whom.

FROM SCIENCE AS PRODUCT TO SCIENCE AS PROCESS

In the last decade or so, however, significant signs of change have begun to emerge in the way museums present science and technology. A number of influences from both within and without museum walls have brought about these shifts in emphasis. For one thing, cinema, television, amusement arcades and heritage parks have all competed for and seemingly won over enough of the museum-going public for some more sober institutions seriously to worry about the chances of their survival at all. And where these forms of entertainment do touch on science and technology, the stories they tell have often seemed far more accessible and lively than the fare on offer in museums. This competition has provided both a spur for museums to try something new and exciting, and quite often even the model of how they might present information and ideas in a more lively fashion.

Another much closer rival attraction for visitors interested in science has been the science centre. Much of the motivation behind these attractions has been the desire to make fundamental scientific principles central to the exhibits, with the implicit criticism that science museums had come to contain remarkably little 'real' science. John Durant has even suggested that 'if museums had been doing their job properly there would have been no need for these science centres' (Durant, 1993: 26). But as Durant has also pointed out, science centres make no effort to relate the scientific principles they explain to any human context at all. Indeed, many do not even attempt to place individual exhibits in larger scientific let alone social contexts (Durant, 1992: 8–10). Nonetheless, they have made it clear that lively and involving exhibitions of science can be extremely popular, and as a consequence have forced many museums to attempt to rejuvenate their style of presentation, if not actually to copy what goes on in science centres.

Another potential source of influence on museum presentation of science has come from the champions of the cause for a better public understanding of science (PUS) – a movement mainly driven by scientists, who, as one of its spokesmen John Durant has put it, were worried about the indifference and even 'general hostility of many towards science and technology'. (Durant, 1993: 26). Although this campaign has in some cases served to concentrate the minds of science curators, it unfortunately suffers badly from a short-sighted, one-sided view of the problem it seeks to redress. Sadly, the PUS lobby sees science as basically unproblematic, with their main concern being focused on an ignorant public who need to be enticed into a proper appreciation of science. For this state of affairs, the media is principally held responsible. The recipe for improvement is curiously similar to what is frequently heard from the Conservative Party these days. The principle objective for scientists is simply to get their message across more clearly, while the message itself it seems is sacrosanct.

Advocates of this point of view tend to have a rather restricted idea about science exhibitions. For them their role is simply to provide factual learning and educational reinforcement of positive science stories. Sometimes, they also admit to an awakening of interest as a secondary benefit (Miles and Tout, 1992: 28). Their view of public understanding does not, however, seem to encompass any sort of critical understanding of how science operates within society. But the role of museums has to be more than this, supplying instead real commentaries on science. And to this end of providing a broader understanding of science as a process, it is difficult to see how helpful the efforts of these 'pure science' proselytizers can be.

A more direct encouragement for museums to examine the processes of science has come from studies in the history, philosophy and sociology of the subject. From the relationship of natural philosophy to politics in restoration England to the influence of the Vatican on the work of Galileo, any number of historical studies have, in the last twenty years, sought to make clear the interdependence of science and the societies sustaining it. The central question pursued throughout this research has been less 'where does our scientific knowledge come from?' and more 'how was it made and how did it become authoritative?'

Amongst philosophers who have more recently attempted to describe the nature of science, a considerable number have argued that the inner workings of science cannot be understood in isolation from its external cultural and social supports. In particular, a whole branch of the subject – one loosely termed the sociology of knowledge – has vigorously argued that scientific ideas are *made* and not simply discovered. Particularly influential here has been the work of Bruno Latour, who has sought to move on from a traditional history of science, which he characterises as being akin to military campaigns recounted by victorious generals, to studies that focus instead on what scientists really do. In his landmark book *Laboratory Life: The Construction of Scientific Facts* Latour described the activities of scientists in a California laboratory in the manner that an anthropologist would an aboriginal village (Latour, 1979; Latour, 1987).

Prompted by these developments, and no doubt others besides, a good number of museums have sought to present their subject not as chunks of information that might be taught in school science classes, but rather as a cultural entity formed in a social context. The examples described below are inevitably selective, standing in for many other projects that I have failed to visit or read about. While none of these projects has been entirely successful, what they collectively indicate is the broad range of strategies available to museums, through which they can begin to introduce audiences to a science that is shown caught in an intricate web of processes.

'SCIENCE IN THE EIGHTEENTH CENTURY – THE KING GEORGE III COLLECTION'

At first sight, the George III gallery of scientific instruments in the Science Museum appears to be a display of nothing but the products of science. Left and right of the gallery entrance are a bust of Isaac Newton and a cast of a medal of the King. Next, on the right in a floor-to-ceiling showcase are two orreries and a 'philosophical table' displayed against a backdrop reproduction of Joseph Wright of Derby's classic picture of scientific entertainment: 'Lecture on the Orrery' (1766). The rest of the spotlessly white rectilinear gallery is seemingly comprised of one clear glass case after another in which are exquisitely displayed an extraordinary collection of scientific instruments. And at one level, the exhibition is as it seems. These instruments were made at a time when science was being put on show, and reflecting this motive, the gallery self-consciously parades these wonderful objects. As if to make this motive clear, a label that accompanies a case with Charles Butcher's 1733 Orrery (unfortunately displayed elsewhere in the museum) is straightforwardly headed 'Please enjoy the Orrery'.

One of the realizations that has come out of the study of museums and exhibitions in the last two decades is that objects can never unambiguously be said to speak for themselves – they need to be read (Butler, 1992: ch. 6). In the George III gallery, the mass of instruments is shown on the face of it as a collection of works of art – just as you would expect to seem them over the road in the Victoria and Albert Museum. On the next level they are interpreted through the simple artifice of gathering them under different subject headings such as 'Navigation and Surveying', 'The Natural World' and 'Scientific Lecturing'. Here we are given the first indication that these instruments in fact served different ends that related to intellectual, cultural and commercial divisions extant in eighteenth-century Britain. The individual labels for each object further relate how and by whom the instruments were made, how they were used and the scientific ideas that underlay this use.

Interspersed with the cases of objects is another layer of interpretation relayed by video presentations. In one programme some of the beautiful but lifeless objects visitors see around them are set in motion: orrery spheres turn, bells move soundlessly in vacuo and an archimedean screw lifts its load. Curiously however, there is never any evidence of human operators: they appear to be true automata. In another video headed 'Science and Learning' audiences are told how eighteenth-century science became a serious part of the entertainment business, with the likes of scientific lecturers going on extensive tours. A third programme describes the emergent formation of scientific disciplines and the relation of scientific developments with the social upheavals brought on by industrialization.

Along with these video programmes, a separate room has been

set aside at the start of the gallery where an audio-visual programme provides, in broad brush strokes, the social, industrial and intellectual contexts for the rest of the display. Here visitors find out about rural poverty, infant mortality, Watt's early steam engine, foreign exports, coffee houses, scientific lectures, instrument-makers and finally the collection commissioned by George III on show in the gallery.

What this gallery manages to do, in what appears to be a rather conventional looking display, is reveal that the objects on show emerged from a variety of intellectual, cultural and social contexts, and that an understanding of those contexts allows one to see reflected in these highly polished, finely crafted instruments some of the processes involved in the creation and dissemination of scientific knowledge in a very specific historical moment: Georgian Britain. The introduction of a variety of audio-visual programmes into the gallery has both captured the movement of some objects displayed and allowed visitors to speculate about what they meant to the eighteenth-century publics who saw them.

The problem with the gallery is that the objects and these broader interpretations presented audio-visually are both physically and conceptually set apart, so that putting them together ends up being harder work than it should be. One fears that many visitors are put off by their first sight of a boring if beautiful looking gallery, without ever uncovering the various attempts to make sense of the instruments in their own terms. Though clearly married in the mind of the curators, the science and its contextual processes seem to have come apart in being translated into the physical details of a gallery plan. The aesthetic cleanliness of the show has been allowed to mask the social and cultural intricacies of what is presented, and so inevitably only allow the gallery a muted impact.

'EMPIRES OF PHYSICS'

Though dealing with a very different period in the history of science, the Whipple Museum's exhibition about late nineteenth-century physics tackles a similar problem of how to suggest the context in which science was made through displays of its instruments. The answer for Jim Bennett and the other curators of 'Empires of Physics' was to think hard about the space in which the exhibition was set up, and to relate it to two very different places in which physics had been formed and disseminated in late nineteenth-century Britain and Germany.

The gallery for temporary displays in the Whipple is located on two floors, and the basic structure of the exhibition attempted to focus on, as it were, different views of same subject – nineteenth-century physics – on these two levels. Downstairs, subtitled 'The Laboratory', physics was presented in the private world of the technician, student and researcher; while upstairs in 'The Exhibition' it was displayed within the public domain of international scientific exhibitions – the world of the entrepreneur, instrument-maker and showman (Bennett *et al.*, 1993; Bennett, forthcoming).

More specifically, the space recreated downstairs was that of the Cavendish Laboratory and the Cambridge Scientific Instrument Company. Shown here were mock-ups of classes that were part of the experimental education introduced in the teaching laboratories of the University; while the optical models, telescopes and electromagnetic induction models were variously displayed on work benches or in showcases in a fashion redolent of the instrument-makers' stock cupboards. The fact that this was a private world dictated that it should be somewhat alien. And as a consequence, little concession was made to visitors who might for example be expecting labels. Instead, they were meant to feel like uninitiated tourists in a foreign country; and what they saw – namely, uninterpreted pieces of equipment – was meant to be just as difficult and uncompromising as the physics produced in the original environment.

A climb up the stairs at the back of the gallery revealed that this was only half the story. Having witnessed science in the making below, visitors were here introduced to the products of that creation, now displayed in the context 'science and technology' as it was presented at international exhibitions. The information here was made more conventionally accessible – exhibits were identified and explained in labels. And the science presented here was deliberately portrayed as finished and unproblematic – exciting new technologies like the telephone and phonograph being quite unapologetically 'over sold' and shown alongside contemporary 'puffing' literature and the medals that exhibitors won at the original exhibitions. The confusion and difficulties witnessed downstairs then had by now disappeared to be replaced by a sense of finished certainty: science in the making had become science made.

The real impact of the exhibition therefore lay less in the success of either floor genuinely to recreate the actual spaces in which nineteenth-century physics was made or displayed than in the comparison between the two – the sense that finished public science rested on the problems and processes of private enquiry, and that the output of the laboratory had relatively little impact unless it reached a forum like the science shows depicted in the second half of the exhibition.

The ideas behind the exhibition were conceptually audacious, and yet lent themselves to being mapped onto the exhibition plan. The problem with the execution of this idea, however, was that for too many visitors the important message about the contrast between the methods of presentation, and implicitly the nature of the knowledge on show, was all but lost through not being made explicit enough. Downstairs, the main difficulty was that the deliberate lack of explanatory material inevitably left many not sure what to make of it all. Expecting an audience to translate its own frustration into a notion that making science is itself difficult was, to be honest, rather naive, and could only have worked if intense tutelage had been provided before visitors started their conceptual journey. Additionally, the idea that both floors were presenting essentially the same science from different perspectives needed to be stated more bluntly, and indeed repeatedly reinforced. The ideas behind the exhibition are explained with admirable clarity in the catalogue; but again it was an over-ambitious expectation that visitors might read through them while standing in the gallery.

The technique of recreating real spaces also raised all sorts of problems, more familiar to the heritage industry perhaps, about how close to reality 'virtual' spaces can be without real smells, sounds and people; let alone the profound intellectual question of how much an *unprepared* visitor could expect to learn even from, time-travel permitting, a real trip to the Cavendish Laboratory in 1890. From this point of view, most visitors would have appreciated the attentions of a live guide. To be fair, a series of public lectures including a re-enactment of a lecture by Maxwell on the subject of the telephone, and workshops for students and school parties based on the replication of classic experiments provided a certain amount of live interpretation to a good number of visitors.

Even accounting for those who did not benefit from this vital ingredient, 'Empires of Physics' still represents a brave attempt to experiment with using the real space of an exhibition to suggest that science does not come ready made, but has to be both created and presented. It would be my contention that greater efforts to turn these important but subtle points into much more explicit ones, and undoubtedly the availability of larger resources to do so, would have made the experiment even more successful.

SCIENCE AS DRAMA

One way of ensuring that subtle points of interpretation do not languish in unnoticed arrangements of objects or unread text panels is to present ideas through the mouths of real people – lecturers, guides and actors. This manner of interpretation, in part at least, abandons the idea that in museums objects speak louder than words, and instead embraces the notion that the narratives of, in this case science, need to be related more directly.

One example of a science museum that does precisely this is the Alexander Fleming Laboratory Museum. The aims of the display are conventional enough: to tell the story of 'the man and scientist, and the discovery and development of penicillin' – one of 'the greatest medical advances of all times' – trumpets the museum's leaflet. An introductory video and series of text panels lays out in televisual and textbook format the outlines of the story behind the remarkable discovery. It also introduces one of the more irritating features of the museum: the heavy-handed acknowledgement of the display's sponsorship by SmithKline Beecham, which at times almost threatens to sideline the genuine interest of the place.

The core of the visitor's experience comes in the room where Fleming made his discovery. Treated almost in hushed reverence, visitors are encouraged to reflect on the fact that this fascinating episode in scientific history took place in this precise location. The laboratory has in fact had to be put back into a room in St Mary's Hospital that had, since Fleming's departure, been more than once converted. Photographic evidence and a mixture of objects either actually used by Fleming or from that period has enabled the curator and designers realistically to reconstruct Fleming's cluttered small laboratory. In truth there is very little that is genuinely exciting about what visitors see, except for the absolutely crucial fact that they know that this really is the place where he did his work. This is the palpable magic that the museum unapologetically exploits.

What brings the display to life, and what makes the story adhere to the space in which it is set, are a group of volunteer guides who lead visitors in Fleming's footsteps, making them imagine the crucial events that led to his chance observation on 3 September 1928 of what turned out to be penicillin mould killing bacteria in a petri dish on his laboratory bench. The guides are from mixed backgrounds, some are retired nurses, others are housewives, while another is a Justice of the Peace. One or two even have direct links with Fleming and his discovery. The real importance of their presence is, however, to set in motion, figuratively speaking, a display that is conventional and necessarily static. In a very simple way, both by telling the rudiments of a standard story and by sharing their own personal experiences of Fleming and the impact of penicillin, it is this human intervention that sets free the visitor's imagination.

Interestingly, the image of the petri dish used as the museum's logo is, in some ways, the scientific 'product' that has been selected and placed at the centre of the visitor's experience – though the object itself is in fact kept as a national treasure by the British Museum. Just as real, and in this context far more important for the museum, is the illustrated story that the volunteer guides conjure up in a space that is charged with the magic of being the very place where the discovery was originally made. This is the sort of experience that should accompany every blue plaque in London.

In many ways the Fleming Museum is conventional, indeed downright old-fashioned – a shrine to a somewhat romanticized event reconstructed in a small room and introduced by a video that is half background and half pharmaceutical commercial. What makes it worth searching out as one of London's finer attractions off the beaten track – certainly for a science enthusiast – is the fashion in which it brings to life a story of science in the making actually in the site it was originally made.

Two more examples from as far afield as the City Life Museums in Baltimore, USA, and an exhibition on genetics shown at the 1994 Motorshow in Birmingham's NEC serve to indicate the variety of science exhibition contexts in which drama can be put to effective use. At the former, a daunting brief for an initiative that tackled the topic of employment in the science and technology sector that should mainly appeal to audiences of disadvantaged inner-city teenagers was turned into a successful museum project by mixing exhibition displays and drama.

The gallery forming the core of the 'Heroes Just Like You' exhibition was based on reconstructions of various work settings including a 1990s research laboratory. In it were displayed both the sort of scientific equipment one might expect, but also convincing arrays of more personal details: a fish tank, Gary Larson cartoons clipped from a newspaper, post-it notes about up-coming meetings, and the like. Periodically, these static displays – interpreted and labelled in a traditional fashion – were moved aside to turn the gallery space into a stage set for dramatic performances. It was here that local actors Maria Broom and Aaron Parrott appeared as a woman experimenter and a technician ambivalent about his employment hopes and prospects.

Rather than famous, unreachable idols, it was these characters – with aspirations, frustrations and achievements that the young, mostly black audiences could directly relate to – who were the heroes of the play's title. What saved 'Heroes . . .' from being a worthy but dull play was the actors' use of the audience's own slang and attitudes, and their ability to leaven the performance with street-wise humour. This was not great drama, but it did engage an initially hostile teenage audience and further managed to encourage them to think of science as a job of work done by people who in many ways were fairly ordinary, even potentially somewhat like themselves.

Another unusual audience for a science exhibition are the crowds who annually come to admire the latest car models at the motorshow. These were precisely the sort of people that the medical charity the Wellcome Trust sought to engage by taking up some 250 square metres of floor space at the 1994 show and filling it with an exhibition on genetics entitled 'Genes Are Us'. They are also the sort of visitors who, certainly in those numbers, it might take a decade or more to encourage to visit the exhibitions on show at the Trust's headquarters in London.

Thinking that audiences might be tired of cars by the time they reached this stand, the 'Genes' exhibition overwhelmingly presented genetics in a human context. Based largely on interactive exhibits, the show sought in part to convey simple messages about

DNA, inheritance, genetic disorders and the role of genetics in medicine. But the overall objective was also to broach some of the ethical issues that the organisers of the show felt could not be ignored in discussions of this branch of science. As in Baltimore, it was decided that the only effective way of tackling this side of the topic was to introduce live human interpretation in the form of street-theatre style performances. In this way, genetics, with both its potential benefits and perils could be explicitly presented as a human endeavour.

Again, what was important was less the potential 'broadcast quality' of the performances than the fact that difficult ideas and issues could be tackled in an immediately engaging medium, rescuing an area of modern scientific practice that clearly has implications for everyone from the seemingly arcane realms of technical expertise.

The use of drama and human interpreters potentially provides the cheapest, most effective, and certainly the lowest-tech means of getting across quite subtle and sophisticated points about how science is pursued and prosecuted in its social and cultural contexts. The point is not that drama shows up the inherent weaknesses of traditional museum displays made up of objects, labels and recreated spaces, but rather that all these elements can be enormously enhanced with a judiciously employed addition of the human touch (Farmelo, 1992: 45–7). In one fell swoop, actors and interpreters can give visitors an orientation, draw their attention to specific exhibits, evoke the lives of the people who made or used them, speculate about their effects on human lives, and open up all sorts of ethical and moral issues that are so difficult to tackle through static exhibition techniques. In short, drama might just be the most effective way of ensuring that visits to science exhibitions actually stir the emotions and thus become memorable (Bywaters and Richardson, 1993: 30; Kavanagh, 1992: 82).

SCIENCE FOR LIFE

Drama, however, is not the only tool available to modern exhibition makers intent on presenting science alive. Interactive computers are another medium that can be put to work towards the same ends. Another initiative mounted by the Wellcome Trust is a permanent exhibition concerning the ideas and practice of modern medical science housed in its Euston Road headquarters entitled 'Science for Life'. A part of this show of particular interest here is that which tries to explore the nature of scientific research (Ballinger, 1993: 32). This section deliberately sets out to combat the widely held image of scientific research as difficult, dry and remote from the real world, and to give visitors a more realistic idea of what type of human activity most medical enquiry actually is.

At its core is a computer game which, in an invitingly light-hearted manner, asks visitors to see if they can work through their own scientific enquiry and locate the cause of a mysterious 'green spot disease'. The didactic purpose of the game is in part to drum home the rather uninspired and fairly conservative inductive model of scientific enquiry in which information is gathered, whereupon hypotheses are formed and tested. What lifts the programme above this predictable-enough message is the intervention of serendipity. For, along with official reports that reveal vital information, it turns out that choosing to read a local newspaper also provides crucial clues in the scientific mystery. The option of consulting with supervisors also produces a response that most student researchers will recognize: 'Sorry, I'm busy right now'. The game is by no means easy, and the temptation to 'Give Up' is strong. 'Sorry', comes the response if one chooses to do so, 'research can be frustrating'. Even if successful, the player cannot fail to see that doing science requires perseverance, teamwork, and sometimes just plain luck in order to arrive at an answer.

The game is a simple one and the points it makes will for many be relatively unsurprising: scientists are not a breed apart, and in their working lives few of them work as automata uninfluenced by preferences, prejudices and straightforward chance. In making the beginning as well as the end of the scientific process seem prone to all the vagaries that humans normally experience when interacting with their environment, this presentation is, however, remarkably unusual for a museum or exhibition presenting science, and as such is a very welcome contribution to understanding the process of science.

THE POLITICS OF SCIENCE

Good as it is, the computer game that simulates scientific research in the 'Science for Life' exhibition is necessarily restricted to a series of very limited computer menus. Significantly, none of them even hint at the broader contexts that surround any scientific investigation undertaken in the real world: most obviously those provided by departments, disciplines, governments and finally countries. This series of levels represents the ever broader social realms in which science has to be formed and on which its results have their impact. Focusing on science in its relations with each of them almost inevitably introduces elements of controversy. For it thus ceases to be simply a question of how the physical world works, and instead becomes a matter of how groups of people relate to each other – colleagues and rivals, sponsors and governors, reporters, critics and politicians, etc.

As David Hull has described in his book *Science as a Process*, much of the inner workings of the scientific profession are in fact based on a fine balance between co-operation and competition – scientific findings have to be usable by others, and yet are jealously guarded until credit for them has been awarded (Hull, 1990). Moving away from the immediate professional context, the realms of official, public and even historical accountability impose further pressures. All of these tensions are endemic to the whole scientific enterprise; and yet curiously this is still the aspect of science that least frequently finds itself presented in exhibitions, perhaps in part because the projects in which it does are so often bitterly reviled by at least some vocal scientists as being wickedly anti-science.

Nonetheless, the last decade has seen a good number of projects that have focused on the reality of science in the making: on the fact that it is 'dominated by controversy and refutation, and that today's "truth" is probably tomorrow's error' (Parkyn, 1993: 34). Thus, for example, a part of the Science Museum's 'Food for Thought' exhibition included the controversial topic of food poisoning – an area of food science in which the facts are still contested and the politics of research is impossible to hide (Macdonald and Silverstone, 1992: 69–87). However, the difficulty of presenting the nature of the controversy while maintaining a suitable 'balance' approved of by scientific advisers meant that in the end the display concentrated on the uncontested and yet ultimately less interesting aspects of the issue.

Another Science Museum exhibition that, on the face of it, presented a controversial issue was the temporary display entitled 'Passive Smoking'. Even here, after deliberately broaching a topic surrounded by contradictory experts and overt political and commercial pressures, the exhibition failed to reflect the roles of all the players involved, in particular making no reference to the effect on the issue of tobacco advertising and the government's half-hearted attempt to restrict it. As Max Ross has pointed out, these absent voices reduced the display to a rather partial one in which 'economic interest and public (mis) information [was] almost entirely evacuated from the equation'. An innovation that was incorporated in this exhibition, and that it shares with a number of other recent displays, was the inclusion and indeed re-presentation (by means of an interactive computer terminal), of the opinions of the visitors themselves (Ross, forthcoming).

One final project that made an attempt to tackle head-on the politics of science was mounted at the Wellcome Institute for the History of Science. Taking a step back from the revolutionary changes in the actual experiences of childbirth during the twentieth century, 'Birth and Breeding: The Politics of Reproduction in Modern Britain' tried to reflect some of the opposing political and cultural disputes that had surrounded a cluster of issues imbedded in the topic. The parameters for the exhibition were largely set by a number of practical considerations. Since part of the purpose of putting on exhibitions in this location is to showcase 'hidden' parts of the Institute Library's collections, the six showcases in the gallery were used to display six archives relating to prominent individuals and organizations connected with the topic kept in the Library. The choice of who to concentrate on was therefore predetermined, although also inevitably partial. Not making this criteria for selection abundantly clear was one of the bigger mistakes of the project (Arnold *et al.*, 1993).

The archives displayed contained an extraordinary range of material, much of it of considerable visual interest and appeal – for example, Family Planning Association posters, Eugenics Society propaganda, National Birthday Trust Fund flags and stamps, Grantly Dick Read's travel scrapbook and Abortion Law Reform Association leaflets. A remarkable fact about the material displayed was that almost none of it was neutral: each poster, leaflet, letter, book and photograph served some point of view, some argument. For this reason, the exhibition was crowded, some would and indeed did say over-crowded, with assertions and refutations, claims and counter claims and disputes of all sorts. And the aim of the display was not so much to make sure that each side of each dispute was given its fair share of display space, but rather to show a whole segment of biological, medical and social science in a state of permanent upheaval. The subject has lived and will continue to flourish by virtue of arguments carried out in public about both how the science should be done and how it should then be used. And for all of its other failings, this exhibition did make that fact abundantly clear.

These examples of recent projects employing a range of exhibition techniques make it clear that museums are yet again well placed to undertake the task of setting science and its history in motion. By revealing the social and cultural dynamics from which a seemingly lifeless eighteenth century instrument emerged; by employing volunteer guides to tell the story of a remarkable medical discovery in the very place it happened; by having actors bring alive the work of scientists; by simulating the stumbling progress of scientific enquiry through a light-hearted computer game; by presenting direct material evidence of how important controversy and political influence is in science: by all these means, and no doubt very many more, museums can begin to move on from science's finished products to a realistic depiction of its myriad processes: discovery, dissemination, overhaul, reclamation, triumph, refutation, use and abuse. In this way, a science that is alive and kicking will once more be found within museum walls.

BIBLIOGRAPHY

Arnold, Ken (1991) 'Cabinets for the curious: practising science in early modern English museums', PhD Dissertation (History) Princeton University.

Arnold, Ken (forthcoming, 1995) 'Trade, travel and treasures: 17th-century artificial curiosities' in Chloe Chard and Helen Langdon (eds) *Transports: Travel, Pleasure and Imaginative Geography. 1600–1830*. (New Haven: Yale University Press).

Arnold, Ken *et al.* (1993) *Birth and Breeding: The Politics of Reproduction in Modern Britain* (London: Wellcome Trust).

Baldock, Janine, and Brookes, Bill (1993) 'Challenging stereotypes', *Museums Journal*, 93 (October): 31.

Ballinger, Christina (1993) 'One step beyond', *Museums Journal*, 93 (October): 32.

Bennett, Jim *et al.* (1993) *Empires of Physics: A Guide to the Exhibition* (Cambridge: Whipple Museum of History of Science).

Bennett, Jim (forthcoming) 'Can science museums take history seriously?', *Science in Culture*.

Bloom, Joel (1992) 'Science and technology museums face the future' in John Durant (ed.) *Museums and the Public Understanding of Science* (London: Science Museum).

Butler, Stella (1992) *Science and Technology Museums* (Leicester: Leicester University Press).

Bywaters, Jane, and Richardson Pippa (1993) 'Breathing life into exhibits', *Museums Journal*, 93 (October): 30.

Durant, John (ed.) (1992) *Museums and the Public Understanding of Science* (London: Science Museum).

Durant, John (1993) 'Rising to the challenge', *Museums Journal*, 93 (October): 26.

Farmalo, Graham (1992) 'Drama on the galleries' in John Durant (ed.) *Museums and the Public Understanding of Science* (London: Science Museum), pp. 45–50.

Ferriot, Dominique (1992) 'The role of the object in technical museums: The Conservatoire National des Arts et Métiers' in John Durant (ed.) *Museums and the Public Understanding of Science* (London: Science Museum), pp. 79–81.

Findlen, Paula (1994) *Possessing Nature: Museums, Collecting, and Scientific Culture in Early Modern Italy* (Berkley: University of California Press).

Forgan, Sophie (1994) 'The architecture of display: museums, universities and objects in nineteenth-century Britain', *History of Science*, 32: 139–62.

Hull, David (1990) *Science as a Process: The Evolutionary Account of the Social and Conceptual Development of Science* (Chicago: University of Chicago Press).

Kavanagh, Gaynor (1992) 'Dreams and nightmares: science museum provision in Britain' in John Durant (ed.) *Museums and the Public Understanding of Science*, (London: Science Museum), pp. 81–8.

Latour, Bruno (1979) *Laboratory Life: The Construction of Scientific Facts* (London: Sage).

Latour, Bruno (1987) *Science in Action: How to Follow Scientists and Engineers through Society* (Cambridge, MA: Harvard University Press).

Macdonald, Sharon, and Roger Silverstone (1992) 'Science on display: the representation of scientific controversy in museum exhibitions', *Public Understanding of Science*, 1: 69–87.

Miles, Roger and Tout, Alan (1992) 'Exhibitions and the public understanding of science' in John Durant (ed.) *Museums and the Public Understanding of Science* (London: Science Museum), pp. 27–34.

Morton, Alan (1992) 'Tomorrow's yesterdays: science museums and the future' in Robert Lumley (ed.) *The Museum Time-Machine: Putting Cultures on Display* (London: Routledge), pp. 128–43.

Parkyn, Madeleine (1991) 'The portrayal of science in museums: implications for scientific literacy', MA Dissertation (Museum Studies), University of Leicester.

Parkyn, Madeleine (1993) 'Scientific imaging', *Museums Journal*, 93 (October): 29–34.

Pickstone, John V. (1994) ' "Museological science": The place of the analytical/comparative in nineteenth-century science, technology and medicine', *History of Science*, 32: 113–32.

Porter, Gaby (1993) 'Alternative perspectives', *Museums Journal*, 93 (November): 25–6.

Ross, Max (forthcoming) ' "Passive smoking": Controversy at the Science Museum?' *Science in Culture*.

Shapin, Steven (1992) 'Why the public ought to understand science-in-the-making', *Public Understanding of Science*, 1: 27–30.

Tröhler, Ulrich (1993) 'Tracing emotions, concepts and realities in history: the Göttingen Collection of Perinatal Medicine' in Renato G. Mazzolini (ed.) *Non-Verbal Communication in Science prior to 1900* (Florence).

Wagensberg, Jorge (1992) 'Public understanding in a science centre', *Public Understanding of Science*, 1: 31–35.

4

A Conflict of Cultures: Hands-On Science Centres in UK Museums

IAN SIMMONS

BEGINNINGS

Although hands-on science as we know it today originated at the San Francisco Exploratorium in the late 1960s, its impact was not really felt in the UK until the mid 1980s. When it was, the two strands that characterize the modern British hands-on scene immediately made themselves felt, the stand-alone science centre, in the form of the Bristol Exploratory (1987) and the museum gallery centre, first seen in the UK with 'Launch Pad' at the Science Museum and Greens Mill in Nottingham, which opened more or less simultaneously (1986), (Pizzey, 1987: 22 and 128).

Since these beginnings, the British hands-on scene has burgeoned to include nearly 30 different centres, with maybe as many again in the planning stage, and has exerted a considerable degree of influence on museum displays in a wide diversity of areas above and beyond science. The stand alone centres now include major developments like 'Techniquest' in Cardiff and 'Satrosphere' in Aberdeen, while the museum gallery centres now include 'Light on Science' in Birmingham, 'Xperiment' at the Manchester Museum of Science and Industry and 'Science Alive' at Snibston Discovery Park, Leicestershire. Such museums have also begun to integrate hands-on with the interpretation of historic collections which will continue to be one of the primary directions for growth at the museum/hands-on interface.

CONFLICTS

This growth of hands-on within the traditional object-centred sphere of UK museums has not, however, been without its problems. The nature of hands-on exhibits are such that they place very different demands on their creators and operators to those made by object-centred displays. If the standard approach to developing and running an object-centred gallery can be characterised as static and procedural, the approach demanded by a hands-on gallery is dynamic and improvisational. Trying to create a hands-on gallery in an institution experienced only in creating object-centred galleries has, in every case where it has been attempted in Britain, thrown up considerable conflicts between the operational cultures of the two approaches.

Part of the reason for conflicts occurring is the very nature of hands-on science itself. It developed in a fairly fluid and organic way outside the confines of museums, and the end result, which we now recognize as a hands-on science centre, was largely the outcome of what were really a series of pragmatic development decisions made during the creation of the Exploratorium. Had similar goals been aimed at initially in an established museum, a rather different final product would have resulted. An inkling of what might have been could perhaps have been grasped from looking at the recently deceased Children's Gallery at the Science Museum. Little wonder, then, that when museums attempt to create a hands-on centre a degree of conflict occurs: they are attempting to build something which by its very nature requires processes and systems alien to museum practice.

The same problems tend to arise when integrating science-centre style exhibits with museum collections, but this adds another level of complexity to the situation. Here, however, I intend to confine my consideration to the establishment of a separate, Launch Pad style hands-on gallery within a collection-centred museum environment.

DIFFERENT CULTURES OF EXHIBITION DEVELOPMENT

The fundamental difference in cultures goes back to the essential purpose that museums and science centres set for themselves. The Museums Association defines a museum as 'an institution which collects, documents, preserves, exhibits and interprets material evidence and associated information for the public benefit' (Barbour, 1994: 445) and, although there is no formalized definition of this nature for science centres, the goals of the Museum of

Scientific Discovery, Harrisburg, Pennsylvania's science centre, very much embody what most centres see themselves as being about. These include 'To provide a unique science learning centre', 'To stimulate an interest in science', 'To develop positive feelings about museums' and 'To make science education an enjoyable experience' (Gleason, 1987: 102). Where the museums role is concrete and practical, the science centre is placing itself far more in the abstract, affective domain.

The modern museum exhibition development process has been characterized, in a very simplified manner, as a linear approach 'Research – Brief – Design – Contracts – Installation – Education Program' (Quin, 1993). While this does not quite represent the complexity and subtlety of exhibition development, it does highlight the manner in which most museums go about the process, as an essentially compartmentalized exercise in which the baton is passed between several teams who pioneer certain parts of the process, albeit with a series of consultative feedback loops to ensure everything is kept on course.

The hands-on centre, in contrast, has a much looser approach to exhibition development, and while the basics of fitting out the display space, e.g. laying carpets, painting walls, are common to both, and while both might involve various types of evaluation, the core process of science centre development is fundamentally different in nature. This comes about as a result of the different origins of the display types involved in museums and science centres. Standard museum displays are basically no-touch experiences involving objects in cases or some form of static environment along with text, various forms of three-dimensional display and audio-visuals, which at most require visitor interaction to activate them. As such, they come from a design tradition which includes shop-fitting, trade stands and domestic interior design. On the other hand, science centre exhibits are primarily active electromechanical devices whose characteristics are (*a*) real phenomena, not simulated ones form the basis of the exhibit; (*b*) the user can exert control over some significant parameters which affect the behaviour of the exhibit; and (*c*) the exhibit is constructed in a manner which allows creative experimentation (Chabay, 1989: 40). The process of creating something to embody these characteristics requires skills, techniques and approaches alien to those used by the creators of standard museum exhibitions and which are far closer to those used by a light engineer building a prototype of a functional machine, which, after all, is what a science centre exhibit actually is.

Ilan Chabay describes the process his team uses to create hands-on exhibits at the New Curiosity Shop in California,

> The process consists of several steps, beginning with deciding the key concepts relevant to the area of Science or Technology to be communicated, and careful attention to the considerations of science content, engineering and pedagogical value is necessary. An essential part is the testing process. Usually we start by building a very simple version of the basic idea. The crude operating version is tested in our workshop to give us a measure of the feeling and appearance of the phenomena . . . we experiment with the response of the system to control, the magnitude of the effects and ease of observation of changes under various conditions. If the idea is a sound one we design and build a working prototype . . . the exhibits are tested with a variety of users in several settings, we observe the way the users respond to the exhibit . . . we look out for potential or actual breakage and danger to the user. We redesign and rebuild some or all of the exhibit in accord with our observations. Then we test the new version and make it available or modify it further. (Chabay, 1989: 39)

This encapsulates the ideal method of creating a hands-on exhibit, an organic, developmental process based around the actual construction of the exhibit where design, in a formal sense, plays a subsidiary role to engineering processes.

At Techniquest, an industrial designer is part of the exhibit-building team and works closely with exhibit developers in creating exhibits which both conform to the centre's corporate identity and function efficiently as hands-on exhibits. A similar approach is now being adopted at 'Xperiment' in Manchester, while at Snibston Discovery Park the design of the final appearance of the exhibits is highly integrated with the exhibit development process by virtue of the majority of the hands-on workshop team having an art/design background as well as being capable engineers.

This organic, non-linear process of exhibit development has been adopted in virtually all organizations which have successfully created hands-on centres. For example, Nils Hornstrup of Eksperimentarium in Denmark has described their exhibit creation process thus:

> Responsibilities for the production of the exhibits was shared by the academic staff members according to their skills and interests, supplemented with a team of part time researchers. With the very flat organisation of the

> Eksperimentarium it has been possible to maintain short decision making processes and a very high level of shared information throughout the development and production period (Hornstrup 1991: 88).

As Melanie Quin succinctly puts it

> scientific and educational integrity are at risk when instead of adopting the project team approach a more traditional linear chain of command is followed. (Quin, 1993: 22)

When it comes to creating a science centre within a museum institution, especially a long established one with a tradition of successful exhibitions produced in a linear fashion, these two cultures of exhibition development come into conflict. This conflict must be resolved if a museum has any hope of establishing a successful hands-on gallery.

DIFFERENT CULTURES OF GALLERY STAFFING

As with the cultures of exhibition development, the cultures of gallery operation in museums and science centres differ significantly in their origins and direction. Again this has its roots in the differing natures of the material they are primarily intended to deal with. In museums, the heart of the experience is the object, which of itself possesses iconic value as well as an accretion of meaning derived from its origins, use and means of collection, to which is added the interpretation of the object applied by the exhibition curators. Hands-on science centre exhibits on the other hand have a far more abstract core, that of the scientific phenomenon, and interpretation is usually left to the minimum by the exhibition creators. Rather, users are encouraged to make their own interpretations of what they experience. This makes the hands-on exhibit more akin to the shopfitting and texts which present the museum object: they are merely the means by which the subject matter of the exhibition is delivered to the public. In themselves they have no intrinsic value; they are a medium for communication purely and simply.

This difference in fundamental nature is reflected in the culture of staffing and operating the two types of gallery. The culture of museum gallery staffing has developed from the need to provide the significant and often valuable objects on display with a degree of security. While communication and visitor service are now very much the watchwords, the origin of this type of staffing lies in the old-fashioned custodial museum warder. While the police style uniforms and forbidding presence are rare in today's museums, assumptions dating from the days when they held sway are still prevalent in museum culture. Assumptions as to the number of staff per gallery are still based on the custodial model, with museums often allocating one or less per gallery, often with a number of galleries patrolled by peripatetic staff. Likewise, the approach to visitor communication tends to bear the mark of the custodial culture. While staff are now usually equipped to communicate successfully with visitors about the exhibitions they are staffing, and the days when a warder in the Science Museum's Chemistry gallery could reply to a query about the location of the Periodic Table by saying 'Tables are in the V&A, they do furniture' (R. Bracegirdle: personal communication) are long gone, the tendency is to expect staff to be reactive, answering visitor queries, rather than proactive, entering into discussions with visitors and encouraging new approaches to exhibits. This has very much been the case at Snibston Discovery Park, where, although it is a completely new museum, it is part of the long-established Leicestershire County Council Museums Arts and Records Service and consequently took on many of that organization's historic assumptions when it was planned.

Science centres on the other hand did not come with such historical baggage, being recent developments, and because they do not have objects of high intrinsic value to oversee, their gallery staff developed right from the start primarily as communicators. Apart from the requirement to ensure good order and public safety on gallery, gallery staff do not have a security role. This means that they tend to be deployed far more intensively on the galleries where they are working, a ratio of one member of staff to every 8 to 10 exhibits is not uncommon in science centres and they are encouraged to be proactive in their approach to users. It is expected that science centre staff will approach a user who looks as if he or she would benefit from help, encouragement or further information, rather than simply respond to questions. They are also more likely to be tightly integrated with the exhibit building/maintenance team in that the gallery staff are equipped and trained to carry out minor diagnostic and repair work on the exhibits while they are working on gallery.

The integrated, active, communication-based approach of science centres is summed up by Hein, writing of the Exploratorium's Explainer programme. She writes that Explainers

> must know the museum well . . . they are trained intensively . . . .They read whatever written material is available on the exhibits and are encouraged to play with the exhibits and learn from questions asked by the public. They learn from the encounter too, as well as from the entire museum staff. (Hein, 1990: 137)

Indeed, the abstract nature of the science centre is such that stimulating person to person communication can be seen as one of its primary functions. Instead of presenting a physical object for viewing, it presents an ephemeral phenomenon for open-ended investigation and that investigation will often involve a group of people, be it a family or members of a school party discussing among themselves what they are doing and experiencing. This discussion helps people to frame their interpretation of the phenomenon they are experiencing and consider new ways of interrogating the exhibit they are using. As such, interpersonal communication is an important component of developing understanding of the phenomenon embodied in a hands-on exhibit. The explainer's role in all this is partly to act as an informed focalizer for the discussion, answering questions, assisting discovery and suggesting new avenues of investigation which users might find fruitful. This is particularly important because good hands-on exhibits are sufficiently flexible to offer opportunities for discovery to people of a wide range of abilities and expertise. A Bernoulli Blower, for example, might amaze and delight a four-year-old experiencing an unexpected behaviour for the first time, give a GCSE student a chance to come to grips with something only come across in textbooks, or provide a physicist with a puzzle as to why the ball moves about within a seemingly steady airflow. A skilled explainer can assist all these users in their encounter with the exhibit. In this sense the physical discourse of the exhibit itself only serves as the starting point for the user's discovery experience, which is further developed in his dialogue with other users and the explainers to produce an increased understanding of the phenomenon being considered. This locates the hands-on exhibit itself as a jumping-off point for a creative dialogue on the science involved (which, indeed, may be internal within the single user's mind) which much of the development of the understanding of the science actually derives from. The explainer, therefore, has a vital role as a stimulus for this process, not usually held by their museum equivalents. They fine-tune the exhibit's communication to the needs of the particular user.

DIFFERENT CULTURES OF LONG TERM OPERATION

The cultures of museums and science centres diverge in their requirements for successful long-term operation and support as well. This not only encompasses the organization's attitude to the exhibition post-opening, but most crucially its attitude to funding.

When a museum exhibition is created, the concentration of effort, money and staff time is dedicated to putting the exhibition together and getting it open. Once open to the public the exhibition is considered largely finished and, although some funds may be dedicated to troubleshooting and, if lucky, dealing with the wear and tear on displays over the years, very little further investment is made during the exhibition's lifetime. Resources are then reallocated to new exhibitions under development. Certainly, monitoring the condition of objects continues and work may be done on these to ensure their continued well-being, but this is part of the museum's ongoing collection care function which is not tied to specific exhibitions and encompasses objects in store as well as those on display. On the whole, then, investment in galleries in museums is a short-term thing and the expectation is that, once constructed an exhibition should be good for a life of up to ten years in some cases, with minimal further investment and attention.

This approach has an influence on how museums see the staff involved with exhibitions. Curators are involved in planning and development, but usually, once a gallery is open, revert to a primary role, dealing with the collection, while designers and technicians move on to the next exhibition to be developed. Indeed, in an increasing number of museums these roles are no longer represented in the core staffing, but are contracted in for specific exhibitions. This means museum staff relate to exhibitions primarily in a development-oriented way. Once open, they are left to front-of-house staff to care for.

In contrast, while science centres require initial capital or revenue investment and dedicated staff time to construct their exhibitions, just as museums do, this is only the beginning of the investment required. Once a hands-on exhibit is out on the gallery floor, in many ways the period of largest investment is only just starting. Science centre exhibits, being electromechanical devices put under intensive usage in the gallery, need to be kept maintained because components such as switches get used thousand of times a day, even with moderate user throughput. This means they wear out quickly, even when extremely tough. Also, many components available have a design life of perhaps 1,000 hours, which when the museum is open eight hours a day, seven days a week, is expended in just over three months. Unlike static museum displays with their passive communication mode, science centre exhibits only communicate when they are active and working. A peeling object label may be undesirable, but will still communicate its message, a science centre exhibit with a crucial component broken is utterly defunct and mute. It only requires a very small percentage of hands-on exhibits to be conspicuously out of order and public perception is that 'everything is broken'. This will happen with as little as 5 per cent of exhibits disabled (B. Singh: personal communication).

The result of this situation is that once a science centre is open, continual funding support is required in order to provide an adequate degree of maintenance cover, and this needs to increase as the exhibits age because they will need an increasing degree of maintenance as more and more components wear out. If this funding is not forthcoming, a steady decay of exhibits sets in, and one by one the non-working devices have to be taken off gallery leaving the exhibition content defined by a ruthless 'survival of the fittest' situation. This usually leaves the least exciting exhibits on the floor as these tend to be those which are most static and harder to break, or are fairly dull so have not attracted much visitor attention. The money needed to avoid this is far in excess of that which museums normally invest in gallery maintenance, so must be specifically budgeted for when planning a science centre and accepted as an annual revenue burden on the museum.

In addition, this attrition on exhibits usually makes their overall lifetime shorter than static museum exhibitions. By the time an exhibit is three to five years old it has become uneconomic to maintain and needs to be replaced, the implication of this being that unless some form of phased replacement programme is undertaken the museum needs to face the full capital or revenue cost of rebuilding the centre every few years. This of course adds a further annual burden to the cost of running the centre.

In independent centres such as 'Techniquest' or the Bristol Exploratory, there tends to be a permanent programme of exhibit-building, leading to a steady change in the population on the floor and a reserve of exhibits which can be rotated on or off display. They tend to build exhibits quickly and cheaply, refining them while in use on gallery thereby spreading the cost of constructing, maintaining and replacing more evenly.

Whether exhibit-building and maintenance is done this way, or as sporadic bouts of building followed by long periods of just maintaining, there are staffing implications. With a science centre, teams can no longer build, open and move on: there must be a team available for rapid response maintenance, and for a centre with 50 or so exhibits this will occupy at least three technicians full time, because otherwise broken-down exhibits proliferate and user dissatisfaction rises. This team, of course, will also require a manager. Whether the team is the same people who built the exhibits, as at Snibston Discovery Park, or a separate team, as at the Science Museum, it still requires a group of staff to be maintained which are not usually present in a museum context, although they can perhaps be seen as the science centre equivalent of the object conservators. This is particularly overt at Snibston where the two teams are run in parallel with equivalent numbers and management structure.

CONFLICT IN ACTION – OPERATION IN BRITAIN'S MUSEUM SCIENCE CENTRES

Looking at the museum-based science centres which exist in Britain today, it is clear that virtually all of them have taken on board some of the requirements for effective operation of a science centre within a museum, but, at the same time, virtually nowhere does this spread across all areas of operation. In almost every case, limitations to the hands-on process derived from conventional museums thinking have handicapped the development of fully successful hands-on centres.

At the Science Museum, for example, the front-of-house operation has achieved an enviable degree of excellence, but so far this has not been reflected in exhibit-building. While the exhibits in 'Launch Pad' and associated galleries are effective, the process which produced them was unwieldly and problematic, involving ideas being generated by an education team, mechanisms then being developed by engineers and finally being passed on to a design team for packaging. This generated considerable internal conflict and handicapped the development process. Also, once hands-on galleries have been opened at the Science Museum, they tend not to undergo further development, and (although they are maintained) they stagnate and eventually decline, while new hands-on projects elsewhere in the museum take precedence. Recently, however, considerable internal changes have taken place in the team responsible for creating and operating hands-on exhibits in the museum and it remains to be seen whether that will alter this pattern.

At both 'Xperiment!', in the Manchester Museum of Science and Industry, and 'Light on Science' in Birmingham's Science and Industry Museum, the requirement to have an effective in-house exhibit building/maintenance team and a full gallery staff has been met by combining the roles into one explainer/fabricator post. This compromise has been fairly successful, allowing exhibits to be built to open the centres in a moderately trouble-free fashion and an adequate level of staff to be maintained on gallery, but it has meant that the museums have had to exclude some gifted communicators or exhibit-builders from their staff as they were not able to do both jobs. Neither have they, however, been able to sustain a high level of long-term exhibit replacement, and have had limited maintenance funds. 'Light on Science' has found it particularly difficult to keep ahead, but so far has managed to do so.

Snibston Discovery Park, on the other hand, has been able to develop the exhibit-building side of the project very successfully, supporting a team of three exhibit builder/maintainers, and receiving continued funding for exhibit development, although, to date, only for new projects rather than renewing existing areas. Maintenance funding has so far continued to be supplied at a tolerable level, although there are signs that this may not continue to be the case. On the front-of-house side, however, the museum has failed to take up the explainer challenge, and operations in this sphere continue to function in the 'modified attendant' mode which has served to limit the communicative effectiveness of the museum's hands-on facilities.

At the other two museums which have incorporated science

centres within a traditional framework, Greens Mill in Nottingham and Science Factory in Newcastle, no major concessions have been made to the different operational needs of the centres within the museum structure. Greens Mill has suffered particularly badly, in that the initial exhibits were built by an amateur exhibit-builder with no real appreciation of the needs of the discipline and subsequent ones were done by the museum's design section who also did not have a full grasp of hands-on exhibits. This process has been described by staff involved as 'a nightmare' (D. Plowman: personal communication) and has considerably truncated the centre's potential. Since opening, the centre has also lacked dedicated maintenance support, so that the exhibits which remain on the floor are there through a 'survival of the fittest' process, selecting out any needing significant maintenance, regardless of their communicative performance. At Science Factory, while exhibit building was done by in-house technicians, no provision was made for longer-term maintenance cover above that provided for the rest of the museum (I. Thilthorpe: personal communication), which has meant a higher degree of breakdown than would otherwise be the case. Both centres have been staffed in line with the rest of the museum's galleries, with standard attendant staff who have not been given any extra training to accommodate the needs of hands-on communication. This has again meant that the communicative opportunities the centre offers have been more limited than might otherwise be the case.

AVOIDING CONFLICT

The conflict between the cultures of museums and science centres is not an inevitable consequence of trying to combine the functions of the two in the same institution. In almost every case limitations to the success of museum-based science centres derive from the inappropriate application of museum methods to the development of the centre. This can be traced back to the early planning and research stages of the project where, often, only one element of the staff involved in the development process has been involved in examining the situation at other centres, if this has been done at all.

In organizations where senior management have been the ones to research the development processes, with the results passed down the chain of command, the risk is run that what they deem important may not necessarily be what is required by the people involved in the practical realization of the project. At 'Xperiment!' the member of staff responsible for its creation found the advice given by management 'irrelevant' to the practical needs of the project (I. Russell: personal communication). There is also the risk that in such a situation, the management team might be unwilling to communicate information which indicates that serious alteration of established practice is required and would be more willing to encourage compromised processes in order to fit into financial or political constraints affecting the task. Likewise, if the research is only done by the staff operationally involved in the project, as was the case at Snibston Discovery Park, with information passed up the chain of command, there is the risk that necessary but unpalatable conclusions can be ignored. In every case in the UK it appears that preliminary research was done in a fragmentary manner if at all, so that crucial people in the management structure did not always grasp the importance of introducing new operational practices to maximize the communicative success of the centre under development. This leads to the first of six key points for producing a successful science centre within a museum.

1 *Involve all levels of staff in the preliminary research*
Senior management will have a different perspective on the problems and requirements of developing a centre to the junior staff involved, but both viewpoints are valid when it comes to considering what the museum needs to do to make the centre a success. This means that all the development staff should look at other centres, independent as well as museum based ones, to see what they believe has contributed to their success and what they see as problems. It is then vital to look at the results of this research in an objective way and seriously address the issues raised in order to get a realistic sense of where the museum should be heading. This is only absolutely necessary, though, if no one on the museum staff has experience of developing and operating hands-on projects. Today there are increasing numbers of experienced people in the hands-on field, but little career structure. Recruiting key people from those who have gained experience in existing centres shortens the learning curve considerably.

2 *Look at the museum's long-term ability to meet the needs of a science centre*
There is little point in investing time and money in creating a centre if the museum lacks the funds or staff to maintain and operate it in the long term. If money to do this cannot be guaranteed, or at least hoped for with a degree of certainty, before the project begins, it is unlikely to be a good idea to begin at all. Also, if there is unlikely to be any money for long-term developments to keep the centre fresh, doubts must be cast on the good sense of proceeding.

3 *Build the exhibits properly*
To get adequate hands-on exhibits it is necessary either to set up an adequately staffed, funded and equipped workshop to do it in house, or to allow for a much larger project budget and contract it out to professional hands-on exhibit-builders. Compromising the process compromises the exhibits. It is highly advisable to avoid succumbing to the temptation to let local industry or universities build exhibits for a centre as a form of sponsorship in kind. No matter how good relationships and communication have been between these organizations and the museum, successful exhibits have rarely resulted because of the exhibit-builders' overall lack of understanding of the environment in which the exhibits need to work.

4 *View exhibit staff as a long term investment*
Once built, exhibits have to be maintained, and the best people to do so are those who built them. They understand the mechanism they built and have fewer inhibitions about rebuilding to solve a problem. Even if exhibits are built by contractors, staff time must still be invested in maintenance; someone must still be there with maintaining hands-on as a priority. If they break down they are useless and the public will be dissatisfied. Centres also need to continue to develop to keep fresh and draw visitors. If exhibit staff are shed on opening, this capacity is lost along with the expertise built up in developing the centre.

5 *Provide adequate numbers of specialised gallery staff*
Hands-on exhibits are stimuli for communication and benefit enormously from being supported by staff who are trained and motivated to enhance this. The temptation is strong in museums to use attendants or some visitor friendly variant of them, rather than commit fully to a specialist team. This, however, inevitably

reduces the commitment and concentration of the staff involved to the detriment of the centre.

6 *Do not allow stagnation*
Once the centre is open development needs to continue to maintain the vibrancy of the place and bring visitors back for return visits. Science centres do not have the draw of iconic objects to do this: they rely on communicating ideas and allowing users to experience phenomena, it is the chance to have new experiences every time they come that draws people back.

CONCLUSION

These points may appear common sense, the sort of things all museums should be taking account of in whatever they do, and indeed they are, but in every case that I know of where a museum in Britain has considered creating a hands-on centre to date, one or more of these crucial points has been neglected or ignored. It is not a problem entirely unique to hands-on: parallel situations arise whenever museums embark on a technology-centred means of communication, such as computer information points. Oker has identified many failings of a similar nature made by museums embarking on this type of display as well (Oker, 1991: 272) and similar examples can be found when things like animatronics, talking heads and even tape/slide shows are employed. As museums move into more technological modes of communication, their methods of working are going to need to be re-evaluated and other institutions learned from. Science centres simply represent one of the more extreme forms of technical and operational challenge that museums will be presented with.

BIBLIOGRAPHY

Barbour, S. (ed.) (1994–95) *Museums Yearbook*, (London: Museums Association).

Bearman, D. (ed.) (1991) *Hypermedia and Interactivity in Museums* (Pittsburgh, PA: Archives and Museum Informatics).

Chabay, I. (ed.) (1989) '*Big exhibits from small toys grow (and vice-versa)*' in M. Quinn (ed.) *Sharing Science* (London: COPUS), pp. 39–40.

Gleason, K. (1987) 'Museum of Scientific discovery' in S. Pizzey (ed.) *Interactive*

Science and Technology Centres (London: Science Projects Publishing).

Grinnell, S. (ed.) (1992) *A New Place for Learning Science* (Washington: Association of Science Technology Centres).

Hein, H. (1990) *The Exploratorium: The Museum as Laboratory* (Washington: Smithsonian Institution Press).

Hornstrup, N. (1991) 'The way we did it in Denmark' in S. Grinnell (ed.) *A New Place for Learning Science* (Washington: Association of Science Technology Centres), pp. 89–91.

Oker, J. (1991) 'Reliability of interactive computer exhibits' in D. Bearman (ed.) *Hypermedia and Interactivity in Museums* (Pittsburgh, PA: Archives and Museum Informatics), pp. 81–9.

Pizzey, S. (ed.) (1987) *Interactive Science and Technology Centres* (London: Science Projects Publishing).

Quin, M. (ed.) (1989) *Sharing Science* (London: COPUS).

Quin, M. (1993) '*How to do two things together – how to renew parts of existing museums and the associated organisational obstacles and challenges*', unpublished paper presented at the 1993 ECSITE Conference at Eksperimentarium, Copenhagen.

5

Science and its Stakeholders: The Making of 'Science in American Life'

ARTHUR MOLELLA AND CARLENE STEPHENS

INTRODUCTION: THE HISTORY OF SCIENCE IN A POST-COLD-WAR WORLD

In April 1994, the Smithsonian's National Museum of American History opened a new exhibition on science and society called 'Science in American Life'. Controversy surrounded the exhibit during its development phase and continued to follow it after opening. Details about the nature of that controversy will follow, but, by way of introduction, we first want to provide a broader perspective. To us, the chief curators of the exhibition, the attention 'Science in American Life' has received signals a new public visibility for the history of science and technology. The issue for all of us, as historians in public museums, is how we learn to deal with this increased interest in our work from the outside.

The higher profile for the history of science and technology is on the whole welcome, but it reflects a more general trend in American popular attitudes: a divided and contentious public's awareness of their stake in the telling of American history. In the United States, history is the focus of public debate today more than at any time in recent memory. At the root of this debate, it seems to us, is a mounting national identity crisis – a crisis that began in the late 1960s but heightened significantly during the Reagan years, especially at the end of the Cold War. With the sudden disappearance of the USSR as a common enemy, the veneer of social and cultural unity began to crumble, exposing cracks and fault-lines in the American cultural landscape. We are now in the midst of culture wars, in which opposing factions compete for cultural space and authority (Shor, 1992; Hunter, 1991). More and more, the battle is joined in the name of historical interpretation, with the ultimate prize the ownership of history itself.

Battles over history used to be waged in books; but now they have exploded into the popular arenas of the late twentieth century. In the past year, for example, plans to depict the history of slavery at a proposed – but now abandoned – new Walt Disney theme park near Washington, DC, ignited a major controversy. The cultural wars have also become the museum wars. The Holocaust Museum in Washington stands at the centre of a continuing dispute about whose story merits a central position on the Mall, home to most of the country's national museums. The Smithsonian has also become the scene of historical controversy. A few years ago, 'The West as America', a Smithsonian art exhibit purporting to reinterpret paintings of the American west as documents of imperialism and genocide, was publicly condemned by an eminent American historian as 'perverse' and 'destructive' (Truettner, 1991; Van Dyne, 1994: 97). National politicians joined the fray, threatening cuts to the Smithsonian budget. Every museum curator knows how dramatically the stakes of presenting history have risen in recent years. Striving to justify themselves and their agendas in terms of the past, opposing constituencies advocate their own versions of history and in fact contend that they are the only ones qualified to tell their story. History, once the pursuit of knowledge for its own sake, has become the applied science of the 1990s.

Today, the cultural wars are spreading into the once-tranquil domains of the science and technology museum. Over the past year, a fierce public debate erupted over a forthcoming exhibit at the National Air and Space Museum on the *Enola Gay*, the plane that dropped the atomic bomb on Hiroshima. Veterans groups and their allies confronted curators in a noisy public clash over the justifications for dropping the bomb, accusing the exhibit's curators of being too sympathetic to the Japanese cause. In an unprecedented editorial, the *Washington Post* joined the controversy on the side of the museum's critics, and in another piece in the same newspaper one of its conservative columnists damned the exhibit's message as pernicious 'revisionism', a term that has itself been revised to connote historical mischief (14 August 1994, p. C8; Krauthammer, 19 August 1994, p. A27). The turmoil only quieted down when the Smithsonian agreed to present the plane essentially without context. Surely, our colleagues must pine for the days when their museum was viewed approvingly as a temple of aerospace progress (McMahon, 1981: 281–96).

'Science in American Life' has come in for similar criticisms of revisionism. As curators of the exhibition, we were shocked to see the exhibit described as an example of 'the Trojan Horse at work' – i.e. the bearer of an allegedly anti-scientific message within a public institution dedicated to science and culture (electronic mail communication, 28 June 1994). It was at once exhilarating and daunting to see how issues debated in academic meetings and journals could become positively incendiary once they entered the public domain. One soon realizes just how many stakeholders are out there for public history, even for an esoteric sub-field like the history of science.

For the Smithsonian and for public museums in general at this time in history, these are far from academic questions. With culture increasingly politicized these days, at least on our side of the Atlantic, museums are at risk of losing support from public funding agencies fearful of alienating major constituencies. Smithsonian administrators are already wondering what the battle over the *Enola Gay* portends for future federal support of the Institution's other programmes. Moreover, at a time of shrinking federal budgets for cultural undertakings, museums like the Smithsonian are increasingly dependent on private funding, especially from large industrial corporations. For the art museum, sponsorship is generally unproblematic since corporations rarely have a direct interest in the content of the exhibition. It is quite another matter for the science and technology museum, however, where corporate funding is more often than not linked to product lines. Railroad corporations tend to support transportation exhibits, pharmaceutical companies, medical exhibits, and so on – and they care a great deal about what these exhibits say. The game becomes far more serious when stakeholders become the funders. Directors of private museums have always had to worry about the feelings of their private backers, but we in the public sector are just learning the facts of life.

As these various stakeholders press their case for their own versions of history, we confess that we have found little evidence of sympathy or understanding for the job of the curator. Our task involves walking a difficult line: to present an honest view of the past that avoids taking sides yet still credibly addresses issues that capture the imagination of a contemporary audience. How do we present more than one side of an issue without offending one group or another and finding ourselves an unwilling participant in – perhaps even a casualty of – the culture wars? How do we confront hard questions without alienating our sources of funding? We have found no easy answers to these questions and, in times of cultural polarization, we find there is no one out there eager to help us puzzle through the complexities we face.

What follows is a brief account of our experiences in this regard with 'Science in American Life'. After a brief overview of the exhibition, we will discuss our interactions with various stakeholders during the exhibition's development phase and during the months after opening. We have learned a number of lessons about the sociology of museum exhibits from this experience, and, through this article, we are hoping to stimulate a dialogue with colleagues who are facing similar situations.

OVERVIEW OF 'SCIENCE IN AMERICAN LIFE'

'Science in American Life' is one of the largest exhibitions ever produced at the National Museum of American History as well as the museum's first major display on science and society. The exhibit covers 1,100 square metres and was supported by a grant from the American Chemical Society in the amount of $5.3 million, including three years of staffing and maintenance. The exhibit explores the evolving relationship between science, technology and society over the past 125 years. The presentation is highly interactive. One-quarter of the floor space is devoted to laboratory experiments, mechanical and electronic interactive stations, and other hands-on activities. About a dozen of these are integrated into the exhibit; the rest are concentrated in two science centres. To our knowledge, 'Science in American Life' is the first American exhibition to combine a science centre approach with historical museum display.

The historical presentation is organized around 22 case studies. Our selection criteria differed from those of a traditional history of science display. Rather than selecting highlights of science – great scientists and their discoveries – we chose episodes that exemplified interactions of science and society. Unifying the case studies is an overarching theme – the evolution of Americans' attitudes toward progress. For Americans who lived in the first half of the twentieth century, advances in science and technology meant material and social progress. By century's end, this attitude has yielded to a more complex appraisal of the relationship between science, technology and progress. It has become apparent that not only benefits but also costs, consequences and social responsibility go hand in hand with the study and control of nature. 'Science in American Life' documents such broad themes as the establishment of science in American universities in the late nineteenth century, the growing cultural authority of science after the First World War, and the public questioning of scientific authority after the 1960s. Case studies include chemist Ira Remsen's research laboratory at Johns Hopkins University as the embodiment of his ideal of pure science; participation of American scientists in Progressive Era reforms; the Scopes trial; the New York World's Fair of 1939–40 as a symbol of the mounting cultural authority of science; the Manhattan Project; a 1950s suburban house furnished with synthetic materials; the birth control pill; the Superconducting Super Collider; and the lab that perfected the technique of cloning.

Our presentation gives as much emphasis to the public response to science and technology as to the work of scientists. Moreover, many of our case studies embody 'hot end' issues, historical episodes with a continuing relevance. Our premise is that it is important for all Americans – scientists and non-scientists alike – to understand science as a responsibility of citizenship in a society increasingly dependent on science and technology.

THE STAKEHOLDERS

In the development stage of our exhibition, we encountered four major groups of stakeholders: curators within the museum; elements of the scientific community; historians of science; and what we used to call the general public. In describing our interactions with these groups, we will focus on one particularly contentious issue – the issue of boundaries.

'Science in American Life' adopts a broad definition of science, embracing much of what is commonly distinguished as technology. We deliberately challenged the traditional compartmentalization of science and technology into separate realms. While we had assumed that definitions of science carried a great deal of ideological baggage, we were still surprised at the degree of emotional investment various constituencies had in maintaining sharp boundaries between the scientific and technological enterprises.

Consensus among the curators on the team was not easily achieved on the matter of boundaries. The exhibit team for 'Science in American Life' comprised a diverse group of historians of science and technology as well as social historians. We had a number of spirited battles, often reflecting the parochialisms of our specialities. Other museum curators, not on the team, had a stake in the exhibition as well. Some of the museum's science curators were offended by the seeming conflation of science and technology and would have much preferred a show built around scientific instruments, especially those from their particular collections. They agrued it was the exhibit's responsibility to enlighten the public about the 'true' distinctions between science and technology. Social history curators were baffled by, even politically suspicious of, the science historians' commitment to transcendent scientific ideas, detecting an evasion of social issues and controversy. One social historian complained bitterly that too much emphasis on scientific or technological artifacts was prima-facie evidence of a right-wing agenda. We finally came to closure with agreement that, historically, the boundaries between science and technology are not given but negotiated. Many of our colleagues in the museum, however, remain unimpressed by this argument.

Some scientists emerged as vociferous critics of our broad definition of science. Firm believers in science as an objective, disinterested, essentially non-utilitarian enterprise, they appeared to believe that too much contact with technology and, by way of technology, with society taints science and undermines the ideal of pure science. Even industrial and Defense Department scientists supported this ideal, seemingly oblivious to the irony of their position. A few chemists were especially sensitive to our inclusion of applied fields. They felt they had to overcome two public relations problems: an age-old stigma of being low on the hierarchy of the sciences and a popular identification of chemistry with pollution and bad things in general. The curatorial team was repeatedly urged to deal more with basic research (read 'pure research') and less with the applied end of things. Scientists urged the curators not to implicate fellow scientists in areas of social concern. They maintained that science is neutral, and that it is only 'society' that uses or abuses science.

On these and other issues, we heard from our sponsors – the American Chemical Society – very early in the game. Having 150,000 members, the society's involvement guaranteed the interest of the rest of the scientific community, especially physicists wondering what the chemists were saying about them. A number of months after the exhibit opened, the leadership of the American Physical Society went to the head of the Smithsonian and urged revisions of the exhibition that would place more emphasis on the spiritual value of science and less on applications. These meetings were immediately reported in the scientific media (Macilwain, 1995).

We also discovered that some university-based historians of science wanted to maintain clear distinctions between science and technology, despite all the academic ink spilled on the nuances of the science-technology interrelationship. In fact, some historians of science on our exhibit's advisory committee, especially those whose professional training was originally in the sciences, seemed to share the concern for the tainting of science by technological or social matters. This is another expression of the same loyalties that resulted in very separate societies devoted to the history of science and history of technology.

The public is a fourth stakeholder in this matter of boundaries. Our pre-exhibit surveys confirmed the findings of many studies that the general public doesn't distinguish in kind between science and technology. To them Einstein and Edison occupy the same world. This opinion is not necessarily naive. The public is aware that science affects them in a wide variety of ways that show no respect for boundaries: the public pays for science, benefits from its applications, and sometimes suffers the negative effects of those applications. As curators, we could have taken the approach of 'correcting' public misperceptions of boundaries. But, for better or worse, we chose instead to appreciate the perspectives of our non-specialist constituencies.

In this battle of the boundaries, some artifacts in the exhibition took on symbolic significance. Where we encountered the most opposition by far was with an artifact representing postwar anxieties about the atomic bomb – a family fallout shelter from Fort Wayne, Indiana. Everyone on the exhibit team itself loved the idea of including this artifact. But some scientists claimed that we had made a category error, that the fallout shelter had nothing to do with science and everything to do with public irrationality. At the root of these criticisms, in our view, was a deep-seated concern that the shelter would stand out as a symbol of scientific evil. Regarding it as one of the most evocative artifacts in the show, we argued that its power lay in ambivalent meanings that had shifted over time. We called attention to changing views of the family fallout shelter: after the war it was considered a realistic technical fix to the destructive potential of the bomb. After the development of the hydrogen bomb, such shelters came to look ineffectual. By the time it was donated to the Smithsonian in 1989 with the fall of the Berlin Wall, it had taken on the character of a quaint museum artifact symbolizing a very happy event indeed – the end of the Cold War. It remains one of the most popular artifacts in the exhibition.

Another contested boundary lay between 'legitimate' and 'fraudulent' science. When we first collected examples of cold fusion apparatus for display in the exhibit, it was at the height of scientific euphoria and media hype over Pons' and Fleischmann's claims that they had successfully produced nuclear fusion in a bottle. Chemists congratulated us for including it in the exhibit. After the scandal over cold fusion erupted, however, a few of them sharply criticized its retention. To reinforce its significance, we wrote contextual labels about the relationship between scientists, the public and the media in the late twentieth century. Critics of our position – both scientists and historians of science – insisted that the corrosive effects on science of displaying what was now deemed a 'fraudulent' artifact overrode any conceivable historical justification for presenting it to the public. Suddenly, respectable science had become fraud and the cold fusion apparatus took on negative symbolism. Arguments against presenting the cold fusion episode took an extreme form. When we included a seventh-grader's science fair project on cold fusion to illustrate public reception of the phenomenon, one critic even accused us of embarrassing the youthful investigator and possibly ruining her future.

Since the opening of 'Science in American Life', stakeholders have continued to make their views known in a variety of ways ranging from the mass media to private communications – in newspaper and magazine articles and reviews, in letters, in phone calls. But, at the time we were preparing this article, the greatest amount of attention came from an unexpected quarter. Opinions about the newly opened exhibit began to appear almost immediately on the Internet, the much-hyped 'Information Highway'.

The Internet is the ultimate vehicle of expression for multiple constituencies. Originally developed for use by the US Department of Defense, it has since grown to connect research laboratories, colleges and universities, high schools, businesses, libraries, small businesses and increasingly, through commercial on-line services, unaffiliated individuals – about 32 million users today, most in the US. But Europe, too, is rapidly going on-line. Thanks to the Internet, there's now an electronic soapbox for nearly every constituency. You can subscribe to 'discussion lists' on particular topics which bring together people with common interests, ranging from artificial intelligence to recipes, from politics to sex.

It was on these discussion lists that we found most of the electronic chatter about 'Science in American Life', coming chiefly from the scientific community and from historians of science and technology. For instance, prompting a spate of e-mail was an editorial by Robert L. Park, head of public relations for the American Physical Society and editor of a weekly electronic newsletter called *What's New*. Park visited the exhibition and came away fuming. Bothering him most was what he considered our violation of boundaries:

> 'Science in American Life:' IS Science the God that Failed? . . . The focus is not on discovery, but on the public's changing view of science – a view that is certain to worsen as a result of the exhibit . . . It's all there: mushroom clouds, a family bomb shelter from the 60s, DDT, and CFCs. As you leave the exhibit, there is a sign warning visitors to 'STOP AND THINK! IS GENE THERAPY SAFE?' (electronic mail communication, 17 June 1994)

Park's commentary spread like wildfire on the Internet, eliciting 'amens' from others who had never seen the exhibit. The Internet version of 'Science in American Life' seemed to take on a life and character of its own, eventually looking utterly different from the original version. An electronic straw man materialized to accommodate the prejudices of cyber-surfers. Here is a typical response to Park's provocative editorial: 'This is indeed a disgrace, if true! The Smithsonian's purpose is (or SHOULD be) to preserve our societal relics. Social comment on selected offending technologies/discoveries may be appropriate, but certainly not a wholesale endictment (sic) of science and engineering.' And another: 'If even partially true, it's the worst thing I've read in quite a while.' This was followed by a discussion of an exhibit that the respondent had never visited (electronic mail communications to History of

Technology Discussion List, 22 June 1994, 23 June 1994). Soon the very title of the exhibit even evolved on the Internet to: 'Is Science the God that Failed?' No wonder the scientific community was mad.

These were countered by e-mail defenders of the exhibit's presumed viewpoints, but, unfortunately, some of our advocates were equally uniformed about the actual presentation. Responding to these errors on the Internet proved most difficult, since not even Park knew how far and wide his comments had travelled (electronic mail communication, 8 July 1994). (Since our full audience evaluation is not quite completed as of this writing, we have not been able to account for the views of the seven hundred thousand visitors that have direct knowledge of 'Science in American Life'.)

Park continued his campaign against the exhibition in an article for the *Washington Post*, prompting a response in which Molella challenged what he saw as Park's misreading of the exhibition (Park, 25 September 1994, p. G2; Molella, 16 October 1994, p. G2).

That so many people could express passionate opinions about an exhibit they had never seen was most revealing. Our Internet respondents – most of them scientists – were so wedded to their own versions of history that the facts seemed irrelevant. What clearer evidence could there be of prior commitments to certain points of view?

It was fascinating to see how the 'Information Highway' turned into the misinformation highway. Although the Internet is frequently hailed as the ultimate in democratic communication because there are no barriers, lack of regulation also means there is no way to stop misinformation or know where it might go among the estimated 32 million users of the network. Once our exhibit entered the network, 'public outreach' took on an entirely new meaning as 'Science in American Life' became a ghostly spaceship of an exhibit lost in the ether. We share our experiences with the Internet in the conviction that all museums will soon be dealing with the same phenomenon as their constituencies adapt to electronic modes of communication.

Despite the problems of electronic misinformation, the electronic highway will almost certainly assume a crucial role in the development, dissemination, and evaluation of exhibitions. One obvious way to minimize misinterpretation is to put the exhibits themselves on the Internet, which is beginning to occur. (For a list of exhibitions on line, see the World Wide Web home page, 'Exhibitions and Images' – http:155.187.10.12/fun/exhibits.html) – compiled by Jim Croft – jrc@anbg.gov.au.) On-line communication networks will take exhibitions far beyond the walls of the museum, making virtual visits a reality and dramatically expanding the audiences for museum exhibitions. Whether via the Internet or old-fashioned comment books, feedback from visitors is essential to the exhibit process, helping curators evaluate what works and what doesn't. A museum's ability to respond to such data will be a crucial factor in gaining and keeping audiences. The advent of Internet represents a quantum leap in both the dissemination and feedback process.

The experience of developing 'Science in American Life' has reminded us that there is no such thing as 'the general public'. In a postmodern world, museums must reckon with multiple publics, stakeholding interest groups who will not tolerate a 'master narrative' crafted behind the scenes by 'experts' who exclude their experiences. We have moved to the front-lines of the culture wars, where a passionate debate rages over the ownership of history and where the goal of attaining some sort of objective middle ground seems remote indeed. Although soliciting criticisms from a fractious public in times such as this may seem foolhardy, curators who elect to avoid this step will do so at their own peril.

BIBLIOGRAPHY

Electronic mail messages

17 June 1994: Robert L. Park, *What's New*, issued from American Physical Society address, whatsnew@ap.org, with Park's disclaimer: 'Opinions are the author's and are not necessarily shared by the APS, but they should be.'

22 June 1994, 23 June 1994: History of Technology Discussion List (HTECH-L@SIVM.SI.EDU).

28 June 1994: Gerald Holton to Paul R. Gross, quoted in e-mail message from Paul R. Gross to a University of Virginia colleague, 28 June 1994, with instructions to forward to any colleagues who were unconvinced that science is under attack, as Gross and Norman Levitt have argued in *Higher Superstition: The Academic Left and Its Quarrels with Science* (Baltimore: John Hopkins University Press, 1994). The message was forwarded to us through two other people.

8 July 1994: Arthur Molella's response to Robert L. Park's 17 June 1994 item, in Park, 'Curator Responds to Characterization of Exhibit in *What's New* (WN 17 June 1994),' *What's New*.

Books and periodicals

Hunter, James Davison (1991) *Culture Wars: The Struggle to Define America* (New York: Basic Books).

Krauthammer, Charles (1994) 'World War II, revised, or, how we bombed Japan out of racism and spite', *Washington Post*, 19 August 1994, p. A27.

Macilwain, Colin (1995) 'Smithsonian Heeds Physicists' Complaints,' *Nature*, 374: 207.

McMahon, Michael (1981) 'The romance of technological progress: a critical review of the National Air and Space Museum', *Technology and Culture*, 22: 281–96.

Molella, Arthur (1994) 'Apolitical science: evidence doesn't support a conclusion of bias in the exhibit', *Washington Post*, 16 October, p. G2.

Park, Robert L. (1994) 'Science fiction: the Smithsonian's disparaging look at technological advancement', *Washington Post*, 25 September, p. G2.

Shor, Ira (1992) *Culture Wars: School and Society in the Conservative Restoration, 1969–1984* (Chicago: University of Chicago Press).

Truettner, William, (ed.) (1991) *The West as America: Reinterpreting the Frontier, 1820–1920* (Washington, DC: Smithsonian Institution Press).

Van Dyne, Larry (1994) 'Storming the castle', *The Washingtonian*, August: 56–9, 95–101.

Washington Post, 'Context and the Enola Gay', editorial, 14 August 1994, p. C8.

6

Communication Strategies in Interactive Spaces

GILLIAN THOMAS AND TIM CAULTON

INTRODUCTION

Exhibitions involving interactive exhibits have characteristically focused more on the actions of visitors manipulating the exhibit and less on the text and visual information that is presented to the visitor. This is perhaps the inevitable result of the insistence on a hands-on experiential approach to learning. However, the experience of the visit and the use that a visitor can make of an interactive exhibit are strongly influenced by the surrounding environment. This ensemble of structures, text, images, illustrations and complementary computer programmes can be described as the communication strategy of the exhibition.

This article looks at three different interactive exhibitions and contrasts the apparent or explicit communication strategies that exist and their implications for the roles of visitors. While most interactive exhibitions are created without a specific target audience in mind, family groups with pre-teenage children and school groups tend to predominate. This article pays particular attention to these groups.

WHAT IS THE PLACE OF TEXT IN AN INTERACTIVE EXHIBITION?

The developers of interactive exhibitions set out to provide a welcoming, attractive, informal, comfortable and easily understood environment conducive to discovery learning. Whether this is the holistic, contextual and multi-disciplinary environment of the children's museum (Lewin, 1989), or the thematic interactive gallery of science, art, history or archaeology, visitors are like day-trippers to an unknown town trying to make sense of their surroundings. As well as requiring help to make sense of the visit, they need to be encouraged to become honorary citizens for the day, playing a valuable participatory role (Thomas, 1994). Thus, visitor orientation is a vital component of the communication strategy.

There are four elements to exhibition orientation: geographical orientation guides the visitor, psychological orientation stimulates the right frame of mind, intellectual orientation encourages understanding of content, and conceptual orientation helps develop associated ideas (Belcher, 1991). It follows that if the visitor is oriented geographically and psychologically, the intellectual and conceptual learning process will be facilitated. Language – whether in its written form as text, or in its spoken form as the verbal interaction of the explainer/enabler – has a clear role to play alongside objects, graphic images, models, audio-visual material and computer aids.

Typically, half the visitors to an interactive learning space are adults. Adults play a key role in the educational success of exhibits, by assisting with the difficult task of interpreting, explaining and teaching. Thus, the primary role of the adult is that of enabler, and every effort should be made to ensure their physical comfort (Thomas, 1992). If they are uncomfortable, they will tend to draw children on to other activities. Adults are more likely to have a positive frame of mind if the exhibition is perceived to be designed for children, and text can facilitate this process. It must make it clear what physical activity should take place, otherwise the exhibit is confusing and 'doesn't work'. It should also clearly outline the educational value of the activity and how the adult can enhance learning; otherwise the exhibit is fun, but of limited educational value (Mulberg and Hinton, 1993). Thus, text has a complex role in that it must not only be understandable and appealing to children, but also interesting to adults in its own right, enabling them to discuss exhibits with their children.

The assertion is often made at interactive learning centres that visitors do not read text. Research suggests that this is too simplistic an explanation, and that whilst most families (especially children) interact with exhibits before reading any labels, they do go on to read the text – especially if their initial interaction is unsuccessful (Dierking and Falk, 1994). Families arrive with their own agenda to exhibitions. How they interact with exhibits depends on many factors, such as the sex and age of the group, their prior experience, the type of exhibit, and the point at which the exhibit is encountered during the visit (people are much more likely to read text in the early part of the visit, after the initial orientation phase, but before museum fatigue sets in) (Dierking, 1989).

Families visit exhibitions as a social event, and people select small segments of text and introduce them into their conversations. This piecemeal selection of text underpins the need for a clear framework for the presentation of text, and a simple conversational style of writing so that these interactions are facilitated (McManus, 1989, 1990, 1991). The interpretation framework should consist of a main message for the whole exhibition, which is the single most important idea that we would like the visitor to leave with. Everything else within the exhibition must be consistent with this message. This is followed by a hierarchy of messages in decreasing importance: ones that we feel we must communicate, ones that we feel we should communicate, and ones that we would like to communicate (recognizing that we expect a diminishing number of visitors to receive each level of messages) (Rand, 1993). In an interactive learning space it may be decided for clarity that only the main and primary messages will be communicated.

If the language used in labels within this framework is inappropriate for the visitor, or if they cannot quickly work out what the exhibit is about, then the exhibit is likely to be rated as dull or boring, or even out of order (McManus, 1991). Effective labels for children should be short, use non-technical language, make a limited number of points, use a large simple sans-serif typeface, adhere to a traditional upper-case/lower-case convention, and be black on white to aid reading by people with disabilities (Belcher, 1991; Serrel, 1993; Brooklyn Children's Museum, 1989). Four simple stages will help ensure that labels are effective:

1 The target audience should be clearly defined.
2 The proposed text should be analysed for grammatical content and reading level (most modern word-processors can assist in this process).
3 The proposed text should be shown to teachers specializing in language for the target age group.
4 Lastly, and most importantly, the text should be evaluated with children (preferably with the prototype exhibit and associated graphic images) (Taylor, 1992).

The interpretation framework must provide text to fulfil several goals. Clear directional signage and introductory texts at the entrance to each exhibition will assist geographical orientation and aid psychological orientation. A large clear title for each exhibit will orientate the visitor conceptually, and clear instructions for manipulating the exhibit provide an essential prerequisite for intellectual orientation. This can be further assisted by background information, which can be presented in a hierarchical manner – perhaps in a smaller font size to differentiate it from manipulative information, or perhaps at adult eye-level given that it is most likely to be read by adults. This strategy might also be used to provide additional complementary information for teachers or parents, perhaps suggesting activities that children might want to do within the exhibition, or follow-up activities back at home or school, to reinforce the learning process.

Specific information informing adults what children might be learning can change a bored parent into an interested observer. However, the boundary between informing parents and patronizing them is narrow, and this approach is more evident in the United States than in the UK.

WHAT IS THE PLACE OF GRAPHIC MATERIAL?

Graphics images, like text, play an integral part in the communication strategy, assisting in the interpretation and orientation process. Accessible to non-readers and to speakers of other languages, they none the less have their own unspoken code which gives a message to all visitors. Graphics can be used in a variety of ways:

1 To identity areas or themes.
2 To create an environment.
3 To reinforce a message on a specific exhibit.
4 To give instructions, either on exhibits or for services.

The corporate identity can be used to give an overall framework within which a graphics policy can be developed. This gives both a visual coherence and a basis for decisions about individual graphics, as well as reinforcing the main exhibition message. This is of particular importance in the orientation of the visitor. It does not mean that graphics have to be of a single style or approach throughout an organization, rather that an overall framework for development of graphics is created. Any change from that is the result of an informed decision to fulfil a specific purpose as opposed to *ad hoc* development of individual graphic styles, which may confuse the visitor.

Both in the general geographical orientation and in exhibition interpretation, illustrations designed specifically for children are perceived positively by parents, thereby enhancing psychological orientation. Graphics designed to create an environment for learning can aid conceptual orientation, whilst a pictogram can simplify manipulative instructions, thereby aiding intellectual orientation. The adage 'A picture is worth a 1000 words' is particularly appropriate in this context, but only if the image is clear and simple, relevant, attractive to children, and easily understandable. A complex image or pictogram can potentially add a further barrier to be disentangled by the museum visitor. Graphics have the additional disadvantage that they date more rapidly than text and are more location specific. An attempt to create a style that corresponds to visitors in one region may not accord with tastes in another.

Graphics can also offer pitfalls for the unwary, giving a message that does not accord with stated policies of the organization, for example, on equal opportunities. While text can be monitored relatively easily to ensure that no racial or gender group is either privileged or excluded, graphics present a more difficult problem. This is particularly the case where characters are involved in the storyline of an exhibition. The choice of a single character is rendered very difficult and recourse is often made to an animal or space alien, which may be gender and race free. Even this is not an ideal situation, as characters may be perceived by the public as having a different connotation, or regularly referred to as 'he' or 'she', not 'it' (for example, the asexual Scoot the Robot at Eureka! is frequently assumed to be masculine). Other solutions involve using a group of characters, designed to represent the range of visitors. This can be satisfactory, provided monitoring of the use of these characters is carried out. Research has shown that relatively small changes in drawing style can affect substantial changes in understanding.

In short, graphics can play a positive role in reassuring the visitor and in aiding comprehension. As with text, formative evaluation of images alongside the exhibit prototype is desirable if the exhibit and its surrounding graphics, objects, models, audio-visual material and computer aids are to be interpreted as part of a coherent communication strategy.

LAYOUT AND STRUCTURE OF THE EXHIBITION

The layout and structure of an exhibition is rarely considered as part of the overall communication strategy, but it has a profound effect on how the public perceives the space and the kind of behaviour that is engendered. The layout of the exhibition – whether it has a closely defined space, a clearly laid out path, a one way course, or a freedom to roam – affects how visitors perceive the exhibition message and the organization's position with regard to the status of knowledge. Recent trends in object-rich galleries have provided numerous examples where visitors are enclosed in strict pathways, suspended or fenced in, whereas the objects are no longer in cases. This technique was first used in animal parks, where, in cars or tracked vehicles, visitors could watch animals wandering freely. Now applied to stationary objects, it says more about the status of the visitor than it does about the value of the object.

Parents with children have preferences which are related to the practical considerations concerned with the security of their children. Spaces with a clear entrance and exit, the ability to see their children or at least to know where they are, are of considerable importance.

Exhibitions can also be conceived with other uses of the space in mind. Providing opportunities for school groups to be organized, for explainer/enabler led activities, or for drama to occur, will affect the organization of the space and its overall feel. This may give an impression of vitality, flexibility or disorder, depending on how it is integrated and controlled.

Of equal importance in the feel of the exhibition, and therefore to be considered as part of the communication strategy, is the provision of additional facilities for visitors. The presence or not of seating, the ability for children to see without being lifted up, the provision of facilities for baby changing and feeding, the inclusion of facilities for special needs groups integrated within the spaces, all give messages to visitors about their own importance to the organization.

The structure used within an exhibition, determined by the overall design, also gives a message to the public. The choice of materials and colour, the quality of the finish, the type of lighting, all identify the exhibition as being for a specific sector of the public and determine a specific type of learning approach. Often a wide variety is achieved within the same organization, reinforcing different messages. At the Science Museum, the contrast between the styles, materials, and finish of 'Launch Pad' and the gallery entitled 'Science in the Eighteenth Century: the King George Collection' could not be more marked, or more appropriate, to differentiate between the different target publics. By these means, organizations send messages to their visitors as to the kinds of visitors they expect and the type of behaviour which will be considered appropriate.

These are as yet relatively uncharted areas, where choice is based on personal preference or intuition. The effect of changing colours to appeal to different groups, of rounding shapes, or of softening materials has not been investigated for exhibitions. This has become a skilled science in the development of manufactured products, but pressures of time and limited finance mean that this is a neglected area in the development of exhibits, where a pragmatic approach is usually adopted. Indeed, it is more of an art than a science; the main choice is made with the selection of the designer, on the intuitive basis for the kind of exhibitions developed by that designer in the past, or on their track record with the target public.

DIFFERENT STRATEGIES

It is now nearly ten years since the first interactive spaces opened in Europe, following the Exploratorium's success in San Francisco. The Inventorium at the Cité des Sciences et de l'Industrie at La Villette in Paris was the first full-scale space to be opened, followed several months later by Launch Pad at the Science Museum in London. These two spaces have experienced very different developments. While both have been successful in attracting more and different visitors to museums, the Inventorium has doubled in size and recently reopened as the Cité des Enfants. Launch Pad, after an initial move, has remained refurbished but largely unchanged. Eureka!, opened in 1992, offers a different approach; while much of the subject matter is related to science and technology, the focus of the organization is children as opposed to a specific content.

Comparing the communication strategies of these three spaces indicates different priorities in targeting and in approach to learning.

Launch Pad

When Launch Pad opened in 1986, it was the first interactive space world-wide that was devoted to technology. Following the Exploratorium model for development and presentation, a series of Test Beds had been created to develop exhibits, using an in-house team. Designers were used to provide an adaptable structure to independent exhibits. A separate identity was created for the space, which was conceived as an independent 'brand' in the Science Museum.

Within the space visitors were, and are, free to choose a pathway of their choice and frequently roam or, in the case of younger visitors, run around. Over 40 free-standing exhibits are arranged in the space, not grouped in any content-based form. Several exhibits are placed in a darkened area. Whilst some of the exhibits are no longer available and others have replaced them, the general organization that was apparent at the beginning is still there today. Each exhibit has a title, some have a number, and all are accompanied by simple instructions. Explainers are available to offer help if required.

The titles vary in their nature, many are purely descriptive, such as 'Computer Video', 'Pulleys and Belts', or 'Tipper Trucks'. Some are close to puns aimed at adults, for example, a building block puzzle entitled 'Hangover Problem'. Others may well cause a problem for the non-scientifically trained visitor, for example 'Harmonic Drive' or 'Fluidised Bed'. The titles are written in large black and white text, but use capitals exclusively, which is not recommended for early learners.

All exhibits have clear operating instructions, some of which are accompanied by illustrations. These have all been developed in response to the public's need and have clearly stood the test of time. Some exhibits give explanations for what is happening, others do not, or suggest contacting the explainer to find out more. There is no regular pattern in the grading of information, and some exhibits have none.

A series of files has been developed and is available for consultation by those visitors who would like to know more. Files relating to thirteen exhibits are currently available and these include information on applications, and give technical documentation, as well as repeating the instructions given on the exhibit (often with a longer explanation). Longer descriptions of the process or other articles related to the phenomena are included. Some files indicate links to other exhibits and objects in other areas of the museum, but no link to other exhibits in the space is made. The files have a 'home-made' appearance and do not appear to have received a coherent graphic approach. The lack of a file for each exhibit reduces their use, but it may reflect visitors' demands.

More recently, a new graphic display was designed to enhance the space and add visual stimulus. A wall display with lights incorporates the Science Museum's logo, and sign language symbols for Launch Pad. It does not include the Launch Pad logo.

Launch Pad was not designed specifically for children but has become regarded overwhelmingly as the children's space of the Science Museum, such has been its popularity with younger visitors. It is probably for this reason that some aspects of the communication strategy appear confused, together with the lack of an overall renewal or expansion of the space. It has been developed by a mixture of accretion and attrition and this is visible within the space. It is, however, enormously popular and successful, so while on an intellectual level its message often seems unclear, the welcome it offers and its informality appeal to visitors.

This needs to be reinforced with a renewal of the space, and plans are afoot to redevelop it for the tenth anniversary in 1996. Building on its evident qualities, the team wishes to reinforce the messages by grouping exhibits, making links between them and creating coherent themes, which will offer more in-depth learning and make better links with the collections of the museum and the world outside. The use of explainers, so popular with visitors, will be reinforced, as will a more substantial use of graphics and back-up material. Visitors who have no knowledge of science will have more resources available to make sense of what they see and experience. The delight that so many young visitors have found in the freedom that Launch Pad can offer will not be lost, but a better environment for developing an interest to the full will be created.

Cité des Sciences et de l'Industrie

The Cité des Enfants within the Cité des Sciences at la Villette in

Paris opened in 1993, as an expanded and renewed version of the previous Inventorium. Catering for children up to 12, their parents, carers and teachers, it has been designed as an integral part of the Cité des Sciences, yet has a distinct identity within it. Additional spaces will be opened for teenagers in 1995.

When the Cité des Sciences opened in 1986, a clear identity was created for all activities within the Parc of la Villette, and the Cité's logo, a red square, is part of that corporate identity. This has been retained and developed for all other activities within the Cité. The Cité des Enfants is clearly part of that ensemble, both by its title and by its specific logo, which is a development of the square.

The distinct character of the space is further developed by the use of a family of characters, presented at the entrance of the space and used throughout in graphics. Chosen to represent a racial and gender mix, these characters are used throughout, particularly in the instructional graphics.

A communication strategy has been developed which offers different levels of information. The space is clearly divided into different zones, first by age group (three to six, and six to twelve), and then by themes within each zone. Each zone has a brief statement of content, aimed at adults, to enable them to find out what can be discovered in each zone. A theme consists of a number of interactive exhibits, backed by an information point and additional resources. Within each activity, the distinction between instructions and additional information is made using colour coding. Illustrations are used to reinforce messages, directly on individual exhibits, and to create an overall environment.

This communication strategy is the result of several years of research and development. The initial Inventorium was divided into two spaces and into themes, but did not have a clear identity of its own within the overall Cité corporate identity. Its importance as a brand for the Cité has been recognized in the creation of its new identity. Themes were established at the beginning, and the policy of explaining to adults what they and their children could discover was in place. The amount of additional information and associated graphics was very limited at the beginning, but research with visitors showed that the time spent with the exhibit could be substantially increased by including explanatory graphics within an area. This may be a culture-specific reaction: the average French visitor would be less familiar with the concept of sharing an experience with their children than an Anglo-Saxon equivalent. In need of more reassurance and in a culture more oriented to reading and the use of illustrations, (in *Bandes Dessinées*, for example), additional information presented in this way presented an adult with an opportunity to participate, without necessarily having to use the exhibit, about which he or she may have felt uncomfortable. Without this support, an insecure adult would often hurry a child on to another more recognizable activity. It is possible that the same reaction would not happen elsewhere.

The Cité des Enfants continues its communication strategy in a wide range of educational materials that are closely linked to the activities of the space. Different versions are targeted at relatively narrow age bands and researched in detail. How different types of illustrations can affect the understanding of the content has also been investigated. A range of portable exhibitions, the Inventomobiles, has also been developed. Aimed at schools, these are packed into a van and visit an area with one explainer. Further development of the communication strategy will be seen in the forthcoming spaces for teenagers, where school groups will also be provided with related video material to take to schools.

Eureka! The Museum for Children

Eureka! opened in 1992 for children up to 12 and accompanying adults. It differs from Launch Pad and the Cité des Enfants in that it does not form part of a larger institution. Using bold primary colours and a strong corporate logo throughout the exhibitions and accompanying educational and promotional material, the development team attempted to integrate graphics, text and exhibits into a coherent communication strategy. Large 'child-friendly' images of children, drawn by children's book illustrator Satoshi Kitamura, appear outside the museum and act as an integral part of overall signage. This strategy, unashamedly copied from Museum De Los Ninos in Caracas, Venezuela, is intended to provide a welcoming environment for children and adults as an aid to psychological orientation. Geographic orientation is provided by a comprehensive signage system of hanging panels detailing main exhibition areas and facilities, also illustrated by a Satoshi cartoon. Some – particularly for the boys' and girls' toilets – are particularly humorous, appealing to children and adults alike.

Whilst there is an overall strategy for directional signage, each exhibition within Eureka! required slight adaptation from the model presented above. The largest single exhibition, 'Me and My Body,' presents the most coherent communication strategy. The exhibition is the only one with an orientation area, and exhibits, images and text are given a uniform treatment throughout, despite the wide variety of exhibit types. A central character, Scoot the Robot, appears in various two and three dimensional guises asking children questions about themselves: in effect, putting children in the role of experts on themselves and their bodies. The gallery incorporates a series of short, specific activities which are clear to children and whose learning possibilities are clear to adults (Mulberg and Hinton, 1993).

Each exhibit has a large title, usually expressed in the form of a question, followed by simple graphics and instructions (with the word 'Do' clearly signified in orange). Illustrations incorporating children of both sexes (in equal numbers), of mixed race, and of varying size and physical appearance are provided in an attractive cartoon-style. The gallery is successful because the context for learning is clear, and learning from each individual exhibit has a cumulative effect throughout the exhibition. The Eureka! team was able to draw heavily on existing research into children's understanding of themselves (the Health for Life Project), and adopted a conceptual approach that is successful in health education. The questions that appear on the exhibits are typically questions that children ask themselves.

Supporting information for each exhibit is available for adults at eye-level, whilst background information is provided at key quiet locations throughout the exhibition in a file for interested children and adults. A passport collected at the entrance and with sections to fill in as the visitor walks around the exhibition, helps to maintain interest.

The overall coherency of approach evident in 'Me and My Body' is missing in 'Living and Working Together' and 'Inventing and Creating' (subsequently renamed 'Invent, Create, Communicate'). Living and Working Together presents a number of environments around a town square (a house, shop, bank, garage, post office, factory, and recycle centre) for children to role play and investigate simple technology. Each environment – developed by smaller teams than Me and My Body – has adopted a slightly different approach to its use of graphics and text, although the broad model is similar. The context for learning is much more difficult to comprehend, although each space does have a short orientation panel (with Satoshi cartoon) outlining the possibilities for exploration within.

Text and graphics have three main roles in this exhibition: operational instructions, supporting information, and suggestions for role play. Typically, role play is not facilitated by written text but by the verbal interactions of museum enablers, who are often preoccupied with explaining the function of exhibits. For example, in the shop the enabler present is frequently overseeing the operation of the (real) till rather than stimulating role play (Reeves, 1993).

The diversity of content within Living and Working Together is both a strength and a weakness. The expectation that something special is around the corner, and the small intimate environments, provide great opportunities for learning. However, the difficulty of presenting the weird and wacky alongside the commonplace within a familiar environment, and within a coherent communication strategy, is only partially successfully met.

Invent, Create, Communicate is a more traditional interactive science gallery which lacks the intimate learning spaces of Living and Working Together, or the simple and coherent interpretation strategy of Me and My Body. The exhibition presents a series of opportunities for children to use communication technology, with a strategy of providing an appropriate context for the use of that technology (for example, a desert island for primitive communication or a yacht for distress messages). In addition, simple communication games are suggested which were devised to illustrate the strengths and weaknesses of each communication device. A character, Squawk the Parrot, is used to aid interpretation, but its role is less clear or compelling than that of Scoot the Robot in Me and My Body.

The exhibition is less successful for a number of reasons, one of which is that the use of real technology, such as the fax or videophone, often requires quite detailed operational instructions (even though the equipment has been simplified for museum use). Another difficulty is that the communication devices often require the interaction of two people at a distance apart. Whilst this is a potential strength and clearly an inherent part of the communication process, it is not immediately obvious to visitors. Like in the shop in Living and Working Together, enablers are typically preoccupied with explaining the technology rather than stimulating role play.

In total, Eureka! is successful because it provides an environment for learning that is clearly aimed at a target of children aged 5–12 and their adult helpers. Its overall strategy communicates a message to adults and children that this is a special place for children to learn by discovery. However, within this framework, some elements are more successful than others. The exhibition was developed in less than two years, allowing little opportunity for formal evaluation of more than basic exhibition concepts. Me and My Body is successful because the Eureka! team were able to build upon existing research into children's knowledge of themselves, but Living and Working Together and Invent, Create, Communicate are more experimental. As a result, in 1993 the Eureka! education team began the process of summative evaluation to build on strengths and eliminate weaknesses.

One problem perceived by the team was that the concept of the children's museum was so unusual some family groups did not get the best out of the visit because they lacked the prior knowledge of what was expected of the adult members (whose sole role was often standing back holding the children's coats). Indeed, an evaluative study revealed that although most adults understood the context for learning at the museum, 60 per cent felt that they needed more information about their potential role, on the content of individual exhibits, and more guidance on which exhibits were suitable for children of different ages. A study of visitor flow also revealed that visitors were disorientated geographically, with 67 per cent beginning their visit by turning into the least successful exhibition (and least appropriate for younger children), Invent, Create, Communicate. This research suggested that visitor orientation in general, and geographical orientation in particular, could be significantly improved (Hesketh, 1993).

CONCLUSIONS

Debate in recent years on the effectiveness of interactive exhibits has concentrated on affective rather than cognitive outcomes, and there is a growing tendency to challenge that a hands-on approach is necessarily one that will lead to minds-on (Jackson and Hann, 1994).

Indeed, the activity can be so absorbing in itself that it can be a clear barrier to understanding, and it is quite possible that a confusing message can be communicated just as effectively as the intended one.

Two aspects of exhibit development require further investigation. Firstly, for exhibit developers a pragmatic evaluative approach is required to ensure that the interests of target audiences are taken into account, alongside consideration of the form and materials of the exhibit, as well as its intellectual content and technical reliability. A clear definition of learning objectives for each target audience is essential as a yardstick to measure performance. Secondly, further academic research is needed in a variety of interactive learning spaces to identify whether changes in the form and materials used in exhibit construction, or in the associated text and graphics, might achieve a significant improvement in the time spent at exhibits and in the quality of the interaction taking place.

The effectiveness of interactive exhibits to attract the public and to provide an enjoyable day out, a memorable occasion, is not in doubt. What is questioned is how their educational validity can be enforced.

BIBLIOGRAPHY

Belcher, M. (1991) *Exhibitions in Museums* (Leicester: Leicester University Press).

Brooklyn Children's Museum (1989) *Doing it Right: A Workbook for Improving Exhibit Labels* (Brooklyn Children's Museum; Brooklyn).

Dierking L.D. (1989) 'The family museum experience: implications from research', *Journal of Museum Education*, 14, (2): 9–11.

Dierking, L.D., and Falk, J.H. (1994) 'Family behavior and learning in informal science settings: a review of the research', *Science Education*, 57–72.

Hesketh, A. (1993) 'Eureka! The Museum for Children: visitor orientation and behaviour', unpublished dissertation, University of Birmingham (Ironbridge Institute).

Jackson, R., and Hann, K. (1994) 'Learning from the Science Museum', *Journal of Education in Museums*, 15: 11–13.

Lewin, A.W. (1989) 'Children's museums: a structure for family learning', *Marriage and Family Review*, 13(3–4): 51–73.

McManus, P. (1989) 'What people say and how they think in a science museum' in D. Uzzel (ed.) *Heritage Interpretation vol.2* (London: Belhaven), pp. 156–65.

McManus, P. (1990) 'Watch your language! People do read labels' in B. Serrel (ed.) *What Research says about Learning in Science Museums* (Washington: AST Centers), pp. 4–9.

McManus, P. (1991) 'Towards understanding the needs of museum visitors' in B. and G.D. Lord (eds) *Manual of Museum Planning* (London: HMSO), pp. 35–51.

Mulberg, C., and Hinton, M. (1993) 'The alchemy of play: Eureka! The Museum for Children' in S. Pearce (ed.) *Museums and the Appropriation of Culture* (London: Athlone), pp. 238–43.

Rand, J. (1993) 'Building on your ideas' in S. Bicknell and G. Farmelo (eds) *Museum Visitor Studies in the 90s* (London: Science Museum), pp. 145–9.

Reeves, K.M. (1993) 'A study of the educational value and effectiveness of child-centred interactive exhibits for children in family groups', unpublished dissertation, University of Birmingham (Ironbridge Institute).

Serrel, B. (1993), 'Using behaviour to define the effectiveness of exhibitions' in S. Bicknell and G. Farmelo (eds) *Museum Visitor Studies in the 90s* (London: Science Museum), pp. 140–4.

Taylor, S. (ed.) (1992) *Try It! Improving Exhibits Through Formative Evaluation*, (Washington: AST Centers).

Thomas, G. (1992) 'How Eureka! The Museum for Children responds to visitors' needs' in J. Durant (ed.) *Museums and the Public Understanding of Science* (London: Science Museum), pp. 88–93.

Thomas, G. (1994) 'Why are you playing at washing up again?' Some reasons and methods for developing exhibitions for children, in R. Miles and L. Zavala (eds) *Towards the Museum of the Future* (London: Routledge), pp. 117–31.

7

Mindful Play! or Mindless Learning!: Modes of Exploring Science in Museums

IBRAHIM YAHYA

> I could solve my most complex problems in Physics if I had not given up the ways of thinking common to children at play.
>
> J. Robert Oppenheimer
> (cited in Samples, 1976: 83)

> Albert Einstein called the intuitive or metaphorical (playful) mind a sacred gift. He added that the rational mind was a faithful servant. It is paradoxical that, in the context of modern life, we have begun to worship the servant and defile the divine.
>
> Samples (1976: 26)

INTRODUCTION

The importance of play as a way of thinking and a vehicle for intuitive or metaphorical mind has been recently dealt with by Gardner (1991) in his book *The Unschooled Mind.* The idea and theory of play, especially after 1979, the International Year of Children, influenced not only the formal learning setting but also the informal learning institutions including museums. Collection and education, amongst others, are the most important functions of museums and they have been very recently considered the 'excellence and equity' functions respectively by the American Association of Museums (AAM, 1992). The process of collecting has already been illuminated by Professor Susan Pearce (1992) at the University of Leicester. She refers to 'collecting as play' and compares many aspects of play with collecting. It is interesting that there are many similarities between the seemingly non-serious (not always!) play and serious collecting. This paper will however attempt to shed light on the other function of the museum – education, with particular reference to exploration of science in museums, using a range of theories of play and learning. Before we attempt to do this, it is essential to understand many significant developments relating to education in the course of the growth of science museums.

SCIENCE CENTRE: A TYPE OF SCIENCE MUSEUM

A number of authors (Danilov, 1976; Hudson, 1987; and Baird, 1986) propose that science museums have undergone a number of stages of development since the inception of the first technical museum in Paris in 1796. Though the number of stages varies from three to five, the commonality between them is that there are basically three main concepts that delineate the stages: the first is the science 'museum' in which the exhibitions are object-oriented; the second is the science 'centre' in which the emphasis is exclusively on the idea- or phenomena-based exhibits without any objects; and the third is the science 'centrum', a concept originally introduced by Orchistron and Bhathal (1984) to represent those science museums or science centres which attempt to take advantage of the positive aspects of both science centres and science museums, in other words, combining object-oriented and idea-oriented exhibits (Yahya, 1989).

Though the first science centre, according to some authors (Shortland, 1987; Quin, 1994), is considered to be the Exploratorium (sometime bracketed with the Ontario Science Centre), the concept had previously been introduced in a number of American science museums including the Museum of Science and Industry in Chicago, the Franklin Institute Science Museum in Philadelphia, and the California Museum of Science and Industry in Los Angeles. In fact a number of 'science centres' *per se* had been in existence well before the establishment of the Exploratorium or the Ontario Science Centre, both of which were not established until 1969 (Danilov, 1976). They are the Science Center of Pinellas County (established in 1960); Pacific Science Center (1962); Center for Science and Industry (1964); New York Hall of Science (1966); Fernback Science Center (1967); and Lawrence Hall of Science (1968). The success and popularity of the Exploratorium are beyond doubt mainly due to the dedication and leadership of Frank Oppenheimer but also to the three exhibit cookbooks that made possible the duplication of exhibits in science centres throughout the world.

No doubt that the spirit of Oppenheimer has inspired a range of science centres throughout the world; however, his approach has been criticized for a lack of management, not leadership. His successor pointed out that

> Oppenheimer gathered and inspired a dedicated group of young people to deliver a great deal of creative energy to the project. On the other hand, no platform was established for continuing growth and renewal. The young and dedicated band is older now, has lived in isolation from the world in many ways for most of their lives, and is extremely reluctant to modify any procedure or objective. With no change, the Exploratorium will remain a remarkable institution, but will become itself an artefact. (cited in Danilov, 1989: 158)

This criticism equally applies to all science centres in the UK that follow the 'tutti-frutti' approach of the Exploratorium in which the disjointed interactives are grouped and presented, while remaining unaware of the thematic approach of the Ontario Science Centre in which a theme would be presented integrating objects with interactives. The Exploratorium in San Francisco and the Ontario Science Centre in Toronto have set the standard for two types, not just one as commonly believed, of science centres, namely the 'tutti-frutti' type popularized by the former and the thematic type by the latter. Unfortunately, all of the UK science centres except Snibston Discovery Park (see Yahya, 1994 for a review) follow the 'tutti-frutti' approach. Some of the science centres in continental Europe, America and India follow the thematic 'science centrum' approach. For these reasons, criticisms which are based on the experience of the UK science centres cannot be generalized to all science centres. Therefore, the criticisms should be looked at within this limit.

'ARE THEY PLAYING OR MERELY LEARNING FACTS?' OR 'ARE THEY LEARNING OR MERELY PLAYING?'

During the 1970s and just after, many authors discussed the potential of science learning in the informal setting (Kimche, 1978; and Tressel, 1980). There are also a number of criticisms. The nature and definition of the science museum or the science centre generated vigorous debates, and a number of criticisms were made against the science centre for its lack of objects. Even now, many museum personnel and professional bodies, particularly the British Museum Association, have difficulty in recognizing the science centre as a type of science museum. This ongoing debate creates a binary pair, to use Pearce's term: 'science museum: science centre'. Pearce (1989) has identified a range of binary pairs of which 'leisure: work' is one. This binary pair is important for our discussion as it can be equated with 'play: learning'. This false dichotomy of play or learning is the main reason for many of the criticisms, in spite of the fact that the science centre is very successful with its visitors across the world. In what follows we will look at the criticisms of learning in the science centre or museum. After analysing the process of learning and play that happens in science centres using a range of theories, a 'spiral' model for science exploration in museums is synthesized and presented. In the light of the model, the meaning of exploration, play and learning will be further analysed and discussed in terms of their relationships to each other and changes in the understanding of their meaning in this postmodern or information age.

A number of criticisms have been made in relation to education and entertainment in the museum setting. The most important and pertinent comment dealing with learning and play was made by Shortland (1987). The new breed of science centres have tended to differentiate themselves from museums. Some of them have deliberately avoided the term 'museum' in their names: for example, the Exploratorium, Techniquest, Launch Pad, Exploratory and so on. Of about 30 science centres in the UK (Johnson, 1992), almost all of them, including museums, try to identify themselves as something different from the museum, and thus meticulously follow the Exploratorium approach to deliberately evoke a non-museum image.

Shortland observed four important problematics in the science centre approach to dealing with science learning: (1) visitors do not learn through play or through an 'exploratory situation'; (2) children do not take time to read labels; (3) visitors do not acquire appropriate science through the science centre experience; and (4) that education will be the loser if education and entertainment are combined in a setting. This paper deals mainly with his first criticism, but we will briefly respond to the final three points. His second assertion that children do not read labels, is questionable. McManus (1989) uncovered a revealing fact in her research:

contrary to prevalent professional belief, visitors, including children, do read labels. She contended that even people considered through observation not to read the labels were found in the recording of their conversations to 'echo text' or read the contents of labels. This has proved that the reading of labels by visitors can be unobserved even by systematic observation, let alone casual introspection including that of Shortland. Having found that research evidence does not support Shortland's second point, we now turn to his third point.

Although Shortland suggests further investigations to ascertain whether or not visitors acquire 'appropriate science', it is not clear what he means by 'appropriate' science. A similar assumption in a study by Durant was criticized by Porter (1993). Porter observed that studies measuring the public understanding of science define people in terms of whether or not they know certain facts and 'perpetuates a top-down approach to science learning.' She finds that 'such studies ignore the informal, amateur and lay methods of acquiring scientific knowledge and practising scientific techniques by such people as amateur astronomers, environmental groups and many others.' This means that the studies measuring the public understanding of science often ignore one side of the coin of scientific knowledge (i.e. the synthetic approach practised by the amateur scientists or ecologists) and emphasize the other side of the same coin, (i.e. the analytic approach practised by the professional scientists or physical scientists) (Golley, 1988).

The fourth and the last point made by Shortland, that education would be the loser if it were to be combined with entertainment, has been widely quoted by museum personnel and has provoked them to think seriously about museum education (see, for example, Ames, 1988; and Parkyn, 1993). Ames identifies two forces that guide the museum management. The first is the institutional mission of the museum that may include such things as statements and objectives of museums. The second is the interests and needs of the market, in this case, the museum visitors. Considering education as the museum mission and entertainment as the market force, Ames gives an excellent discourse on museum management. However, MacDonald (1988) demonstrates that the marketing force need not have to be only entertainment; it can also be education. The Disney World's EPCOT centre produces many exhibitions with authentic real objects to foster learning and to give an authentic experience, in responding to the marketing force, even though the mission of the Disneyland-type institutions is entertainment. A postmodern approach to the museum accepts that entertainment and education are equally valid, irrespective of their identification as mission or market.

We will now turn to the first point made by Shortland. Jerry Wellington, a lecturer in the Division of Education in the University of Sheffield responded to Shortland's first criticism, which was that visitors do not learn through play or through an 'exploratory situation', through a series of articles (Wellington, 1989 and 1990). He found that the criticism raised by Shortland had also been echoed by others including many parent-visitors. In one of his articles (1990) he tried to answer the question, 'It's fun, but do they learn?'. He found out from interviewing many teachers that the distinction between playing and learning is a totally false one and that all the teachers felt that the science centre did make a contribution to their pupil's education. He was convinced that the fact

> that children are actually playing and being entertained is not seen as drawback but as an advantage by those involved in education; only those critics who view that [children's playing and being entertained] from an objective and external perspective see a distinction between playing and learning and go on to make this into a problem.

Wellington felt strongly, supported by the evidence of a video-recording of the activities going on inside a science centre, that there is far more to learning science than just understanding. He finally concluded that science education has three basic aims: cognitive, affective and psychomotor, and that the science centre contributes in some way to all three areas. The contribution may be not as great in the cognitive domain but is certainly significant in the other two domains.

A similar debate has been aired between Wymer and Quin in *New Scientist*. Wymer (1991) questioned the purpose of the science centre and, although acknowledging that the activities are enjoyable, how much do these disjointed activities in the science centre help children to learn about science, and what, if any, is the connection with public understanding of science? He also observed that the exhibits in most science centres are concerned mainly with physical sciences, and that their links with everyday life are sparse.

In response, Quin (1991) argued that physics is an obvious starting point for practical reasons. Due to the limit of funds and scope, difficult subjects are rarely a priority. There are however science centres in Europe such as Heureka in Finland which use the thematic approach, unlike British science centres. She went on to observe that visitors to science centres 'are looking for an "edutaining" alternative to cinema, and delighted to find a welcoming place where science is not scary and there are no wild-eyed boffins'. She finally concludes that the reason for funding the interactive science and technology centres is justified by their goal of motivation which is a prerequisite and prime constituent for the public understanding of science.

In her most recent work, Quin (1994) argues that the question 'Are they learning or merely playing?' should be reversed to 'Are they playing – developing an exploratory approach to life itself and the basis perhaps for a career in scientific research – or merely learning facts and figures?' This reversal reflects the importance of play in all walks of life, as demonstrated by Kubie's work:

> The free play of preconscious processes accomplishes two goals concurrently: it supplies an endless stream of old data rearranged into new combinations of wholes and fragments on grounds of analogic elements; and it exercises a continuous selective influence not only on free associations, but also minutiae of living, thinking, walking, talking, dreaming, and indeed of every moment of life. (Kubie, 1971: 39)

Quin further goes on to answer herself that the exhibits are only the tip of the iceberg of science communication. Science centres offer a number of special events and programmes more than just exhibits: in-house educational programmes; outreach programmes; science gardens; laboratory science; audio-visual or multi-media shows; lecture demonstrations; science drama; planetarium shows, and so on.

It may be argued that entertainment and education can be effectively combined in the science centre to generate playful and learning activities, both of which are essential to make the experience enjoyable and at the same time useful. To provide the visitor with a memorable and enriching experience, science centres in the postmodern age have already started to combine a range of false dichotomies: object-oriented and idea-oriented exhibits; object-centred and people-centred activities; 'disjointed' and 'thematic' exhibit approaches; and a number of others. Having discussed a range of criticisms and counter arguments and pondered over the importance of entertainment and education, we should now look at theories of play and learning.

THEORIES OF PLAY

Johan Huizinga, a historian, attacked the way psychologists and sociologists looked at play: all of them assumed that play must serve some purpose which is not play, that it must serve some kind of biological purpose. In his classic *Homo Ludens* (Huizinga, 1951) he argued that 'play is more than a mere physiological or a psychological reflex'. It is a function of culture, one of the main bases of civilization, not associated with any particular stage of civilization, but a universal and integral part of life, human as well as animal. Play is central not only to the development of children but also to adults and to the whole community and society.

Play is undoubtedly a mechanism by which humans and animals explore a variety of experiences in different situations for diverse purposes. For example, a good number of people who buy new equipment such as video or washing machines would fail to read the instruction manual; they rather prefer 'to play' with the controls and functions. This demonstrates that individuals come to terms with novelty or familiarize themselves with new things through 'hands-on' experience. This experience of a real situation with a real purpose normally leads to the intuitive understanding of the operations of instruments and machines, which is reinforced subsequently by reference to the manual and consolidated by practice.

People work at their play or play at their work: for example, work in order to win a tennis match; work for the satisfaction of completing a cardigan; writing poetry for the sheer love of words or painting for its own sake; and so on. In her book, Moyles (1989) presents three different forms of play in school, namely *physical play*, *intellectual play*, and *social/emotional play*. Play can therefore demand a range of skills and knowledge, not only physical but also intellectual and social. Applying Norman's theory of complex learning to child's play, Moyles argued that play is a spiral process involving the three processes namely accretion, restructuring and tuning, which we will see later in a greater detail.

The theories of play, according to Cohen (1993), tend to fall into one of the three traditions: Piagetian tradition; Freudian tradition; and educational traditions after Froebel and Montessori. We will not discuss them all in detail as it is beyond the scope of this paper, but will concentrate on the Piagetian tradition. Piaget's theory of play is the most important one for our discourse because of its basic premise that 'to know or to understand is to transform reality and to assimilate it to schemes of transformations'. According to this premise, interaction and doing things are the most important factors in the learning situation. Real learning and understanding is possible only when a person is allowed to manipulate and interact with the environment. This active process of learning is the basis of the science centre interactives or hands-on activities.

Piaget's contribution to play is mainly an outcome of the observations of his children that is discussed in his work *Play, Dreams and Imitation in Childhood* (Piaget, 1951). His theory of play is mostly based on his general theory of cognitive and intellectual development. Before turning to what Piaget means by play, it is essential to look at some key aspects of his theory of cognitive and intellectual development. According to him, children undergo developmentally four stages that are determined genetically: sensory motor, pre-operational, concrete operational, and formal operational. The stages are very much related to the physiological age of children: 0–2 years for the sensory motor; 2–7 years for the pre-operational stage; 7–11 years for the concrete operational stage; and 11–15 years and onwards for the formal operational stage. Though there are research findings that do not support the exactness of the years, there is a general consensus of agreement for his stages.

For Piaget, a fundamental concept underlying all the stages is adaptation. It consists of two very important processes which every child or adult undergoes simultaneously – *accommodation* and *assimilation*. Accommodation is the process by which a child or adult accepts that external conflict in ideas is due to a deficiency in his or her own internal mental structure (which Piaget called schemata) and changes it. Assimilation is the process by which a child internalizes external observations and fits them into his or her schemata. It is the dynamic equilibrium between the two fundamental processes that keeps the organism (the child or adult) developing (Bagchi, Yahya and Cole, 1992).

According to Piaget, 'play' is a sort of state in which *assimilation* of external reality to pre-existing concepts is dominant over *accommodation*; when *accommodation* to external reality is dominant over *assimilation*, he called it 'imitation'. Since assimilation is an aspect of all behaviours, every behaviour has at least some play-like aspects. Therefore, there are only more or less playful behaviours rather than play or non-play behaviours (Gilmore, 1985). Piaget's theory of play gives rise to three broad categories of play: practice play; symbolic play; and games with rules. As their names suggest, the categories proceed from mere practice to complex play with many rules which require learning of symbols. In spite of its cognitive reductionism, Piaget's theory of play has been a sound theory to date and it reveals many aspects of play. Sutton-Smith (1971) adds an interesting insight to Piaget's theory of play:

> Play is not solely a cognitive function (nor solely affective or connotative), but an expressive form *sui generis* with its own unique purpose on the human scene. It does not subserve 'adaptive' thought as Piaget defines it (though of course it can do that); it serves to express personal meanings. It is therefore remarkable in cognitive terms for its uniqueness, in affective terms for its personal expression of feelings, and in connotative terms for its autonomy. (Sutton-Smith, 1971: 341)

THEORIES OF LEARNING

Dr Eilean Hooper-Greenhill, University of Leicester, excellently captures the changing nature of education and learning in museums; science centres are no exceptions:

> it must not be forgotten that in recent years, education itself has become closer to leisure. Progressive educational theory has always maintained that we learn while we are involved, committed and enjoying ourselves. The recent stress on learning through doing tasks in the real world, through links with industry, through visits to sites such as shopping centres, museums and galleries and the stress on course-work instead of exams, all have the effect of making the educational process closer to and more open to other social processes. This fundamental change in ways of learning and teaching means that learners no longer think that they must suffer to learn, and that learning has to be difficult to be effective. (Hooper-Greenhill, 1994: 114)

With the above preamble, we will now introduce a few theories of learning. There are innumerable theories of learning from three main traditions: behaviouristic, cognitive and humanistic. Each of them has different assumptions about the nature of the learner and the learning process. One interesting theory of complex learning suggests many interesting ideas on the nature of play (Norman, 1978). According to Norman, there are three different processes which essentially link a learner's present knowledge with new experiences to acquire new learning: *accretion, restructuring*, and *tuning*. *Accretion* is the process by which the individual acquires information to fill in the mental structure (Fig 7.1). This more or less resembles Piaget's *assimilation*. *Restructuring* is the process by which the learner enhances his or her mental structure to understand the acquired knowledge, i.e. making new concepts and understanding, and resembles Piaget's *accommodation*. *Tuning* is the process by which the learner begins to master the performance or learning with continued use. Ultimately, he or she reaches a stage of automation in the domain through practice and problem solving.

This theory accounts very well for what is going on in the science centre. For example, young children, through exploratory play, spend a great deal of time in the process of *accretion*. They accumulate a set of discrete notions about a particular material or activity but, once this becomes familiar, the child will increasingly be able to perceive underlying patterns or concepts and to begin the process of *restructuring*, sometimes with help from parents or explainers. This process is then likely to be followed by a new period of *accretion*. The cycle is likely to be repeated and operated until a period of *tuning* emerges in which a new learning experience has been thoroughly acquired and becomes '*automatic*'. This period is characterized by a fluency or mastery of these recently acquired concepts or skills.

The modes of learning – accretion, restructuring and tuning – do not necessarily occur in sequence but they are always present as represented in Figure 7.1. The sweep of the radius of the circle represents the nature of learning that has neither a definite starting point nor a definite ending point. The start always builds upon previously acquired material. Learning is therefore not a unitary process and there are different modes of learning, each with different behavioural and instructional assumptions. Having examined the main thrust of this theory of complex learning, we now turn to writers who assume play as a motivational factor in learning.

Motivation is a very important factor in the learning situation. The goal of motivation is considered to be the reason for the public

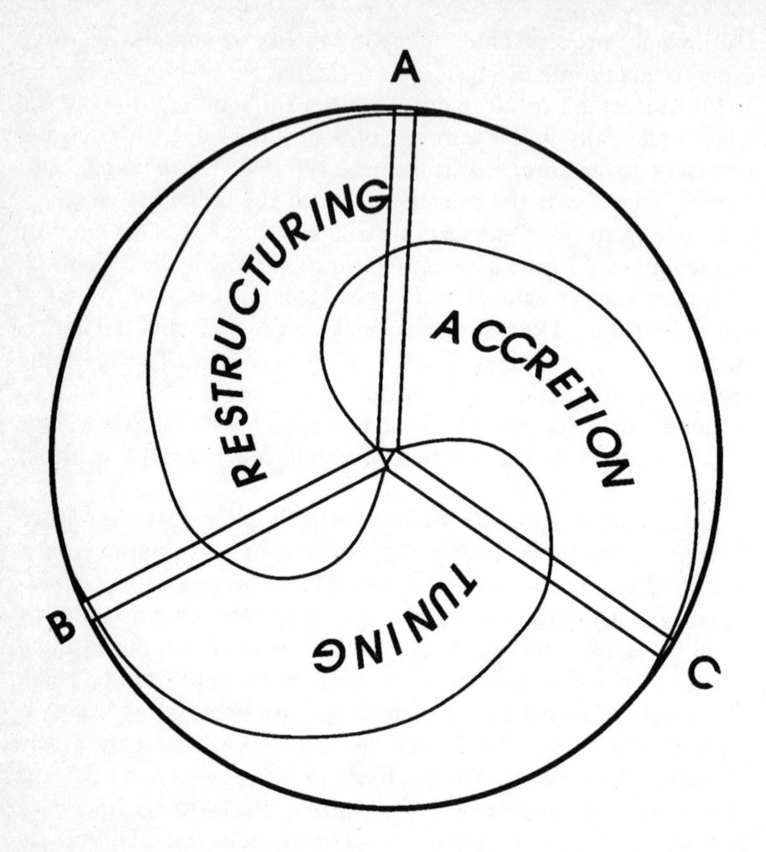

Figure 7.1: Nature of action of three modes of complex learning. A: the radius represents a situation when more of *accretion* followed by restructuring and less of tuning takes place. B: The radius represents a situation when more of *restructuring* followed by tuning and less of accretion takes place. C: the radius represents a situation when more of *tuning* followed by accretion and less of restructuring takes place. The circle symbolizes that there can be any one of the situation (A, B, C, or *any other*) to start with and that an individual can move in either direction (clockwise or anti-clockwise) depending upon one's past experiences.

funding of science centres (Quin, 1991). There are a number of motivational theories of learning that often attempt to emphasize the advantages of combining play with learning or combining entertainment with education as motivational factors. In what follows, we will introduce motivation and review two ways of explaining motivation.

Motivation generally derives from two sources. If motivation is influenced by the internal factors such as satisfaction or enjoyment, then it is called intrinsic motivation. On the other hand, if it is influenced by external factors such as grades, points, recognition of status, or money, then it is called extrinsic motivation. Behavioural psychologists tend to emphasize the importance of extrinsic motivation, whereas cognitive and humanistic psychologists tend to emphasize intrinsic motivation. In reality, both motivations are applicable in any learning situation, but one form may be more appropriate in a particular situation than the other.

The work of Csikszentmihalyi and Robinson (1975) is important in understanding motivation. They used questionnaires and interviews to identify intrinsic rewards, using a group of people – chess players, surgeons, rock climbers, dancers, music composers, and basketball players – who were deeply involved in activities which required much time and effort and skill yet produced little or no financial status. They found out that people who enjoy what they are doing, enter a state of 'flow': they concentrate their attention on a limited field of stimulus, forget personal problems, lose their sense of time and of themselves, feel competent and in control, and have a sense of harmony and union with their surroundings.

The analysis in the above studies revealed a theoretical model of enjoyment as shown in Figure 7.2. According to this model, a person has action capabilities, that is, *skills*, and the activity poses opportunities, that is, *challenges*. When a person believes that the challenges are too demanding for his or her skills, the resulting stress is experienced as *anxiety*; this anxiety reduces to *worry* if the level of challenges is reduced but still higher to face with the existing skills of the person. The state of *flow* is felt only when the challenges are in balance with the person's skills; the experience is autotelic. When skills are greater than the challenges posed, a state of *boredom* results; this state again fades into *anxiety* when the skills are too much as compared to the challenges.

An important outcome from the above study is a clear understanding of the dichotomy between play and work. Generally, rather traditionally, work and play are considered to be diametrically opposite activities. The evidences from the study

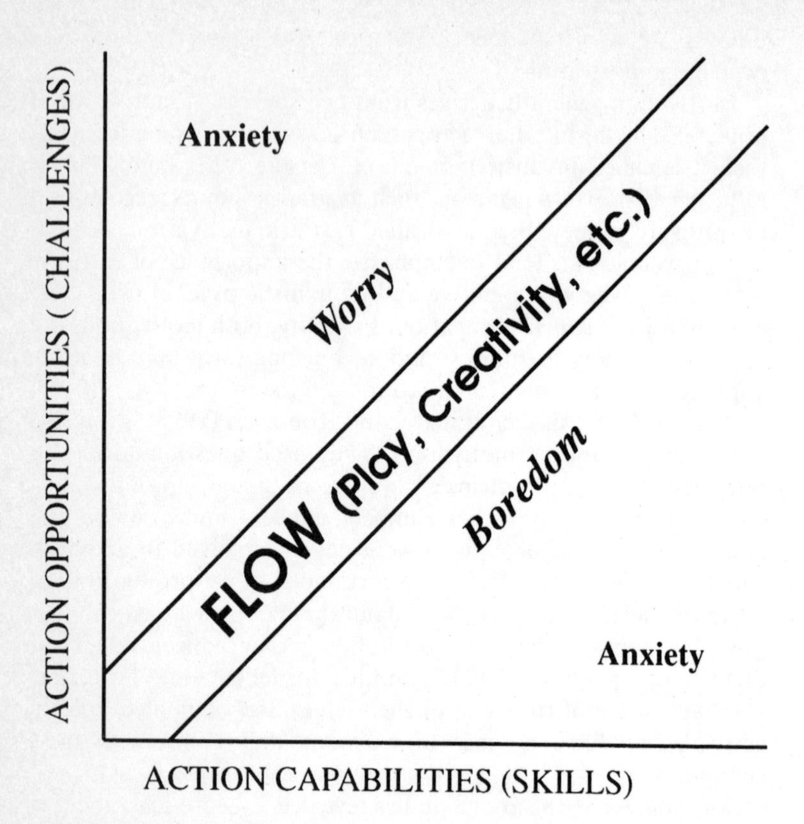

Figure 7.2: A theoretical model for enjoyment (after Csikszentmihalyi and Robinson, 1975).

suggest that the essential difference is not between play and work as culturally defined activities but between the 'flow' experience – which typically occurs in play activities but occurs in work as well – and the experience of worry or boredom, which may occur any time and at any place but is more likely to happen in activities that provide either too few or too many opportunities for action. They conclude that flow experience is the only important factor and where it happens does not matter:

> Work is not necessarily more important than play and play is not necessarily more enjoyable than work. What is both important and enjoyable is that a person acts with the fullness of his or her abilities in a setting where the

> challenges stimulate growth of new abilities. Whether the setting is work or play, productive or recreational, does not matter. Both are equally productive if they make a person experience flow. (Csikszentmihalyi and Robinson, 1975: 202)

The second motivational theory is that of Thomas W. Malone (1980). He conducted a controlled experimental study to find out the factors that make learning fun, using one out of 25 popular computer games, which he surveyed using 65 elementary school students who have been using the games from 2 to 24 months. With assistance from the developers of the software, he developed eight versions of the game with various features deleted from the original version for the experimental study. Eighty fifth-graders were assigned to one of the eight versions. From the results, he found out that there are three factors that affect learning and make it fun: *challenge, fantasy*, and *curiosity*. He, along with Lepper (Malone and Lepper, 1987), presented a taxonomy of intrinsic motivations for learning. To Malone's original 'individual' motivations – *challenge, fantasy, curiosity* were added a fourth class of 'individual' motivations – *control*, and three 'interpersonal' motivations – *co-operation, competition, recognition*.

This taxonomy was adapted by Deborah Lee Perry (1989) to a museum setting. A particularly popular exhibit 'The Colour Connection: Mixing Coloured Lights' was used. She identified six components of an intrinsically motivating museum experience: *curiosity, confidence, challenge, control, play, communication*. The research was intended to develop a model for designing educational museum exhibits by increasing both the visitor's interaction with the exhibit and their social interaction. She set up an interrupted time-series quasi-experiment designed to verify the educational effectiveness of the exhibit and the usefulness of the model with the *existing exhibit* and the newly constructed exhibit (*revised exhibit*) which incorporated the guidelines of the model. The result of this study has indicated that visitors enjoyed both exhibits (*existing* and *revised exhibit*) the same amount but they stayed at the *revised* exhibit for longer periods of time; the amount and quality of interaction with the revised exhibit were increased; the amount and quality of the social interaction were increased in the revised exhibit; and teaching and learning behaviour occurrences were more at the revised exhibit.

Having further tested this model at SciTrek in Atlanta, Perry

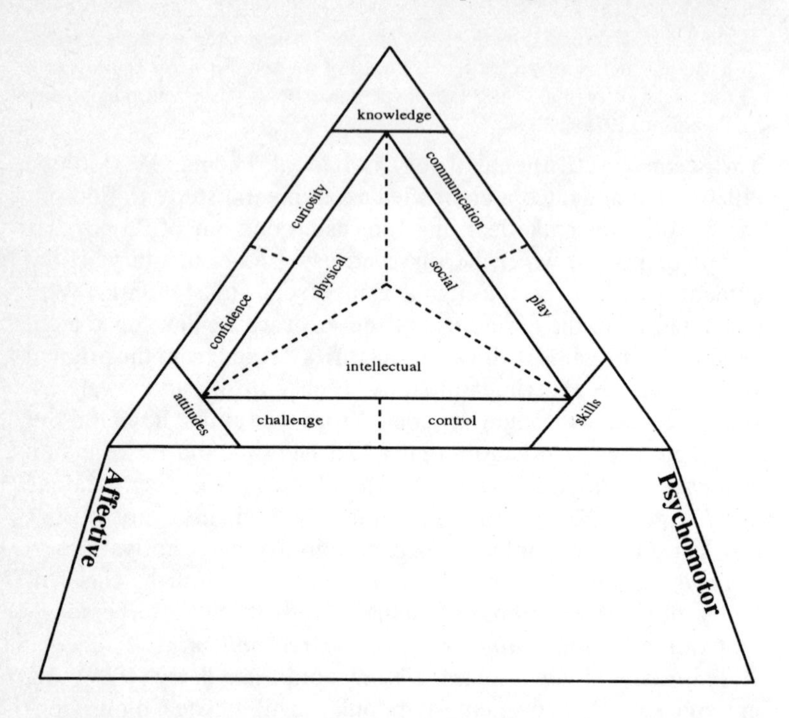

Figure 7.3: A rearranged form of the Anatomy of a Museum Visit originally proposed by Perry (1993). The inner triangle represents *interactions* between the visitor and museum in three different levels. They resemble Falk and Dierking's contexts or Annis's spaces of the museum visit. The corners of the outer triangle represent the three *outcomes* from the museum visit. These are the representatives for what go on in the three knowledge domains or areas in the 'spiral' model. Between the *interactions* and *outcomes*, there are six *needs* of the visitor. The entire form in its triangular shape can be viewed as a cross-section of the knowledge tetrahedron in the 'spiral' model for the exploration of science in museums.

(1993) most recently proposed a 12-component model on anatomy of a museum visit incorporating the earlier studies (see Figure 7.3 for a rearranged form of the model). According to her, these 12 components constitute three factors that affect visitors' learning or broadly, experience in museums: they are *interactions* between the visitor and the museum; *needs* of the visitor; and *outcomes* of the museum visit. The *interactions* between visitors and museums can take place in three different levels. They are in a *physical* level such things as pushing buttons, walking onto galleries, reading labels and so on; they can be in a *social* level such things as explaining and discussing with group members; they, in an *intellectual* level, include such things as thinking on and finding new information. These levels more or less resemble the three 'contexts' proposed by Falk and Dierking (1992) and the three 'spaces' proposed by Annis (1986). Identifying six psychological *needs* of museum visitors, Perry (1993) argues that they have to be met for meaningful learning to occur in a museum environment. They are *curiosity, confidence, challenge, control, play*, and *communication*. Arguing that visitors must be changed in some way in order for a museum visit to be successful, she presents three *outcomes*, namely *knowledge*, *attitudes* and *skills*. These, more or less, resemble the cognitive, affective and psychomotor domains of learning respectively.

The colour connection exhibit in La Villette, was also studied by Lucas *et al.* (1986) in four versions. This exhibit is supposed to demonstrate mixing of colour using three coloured (red, green and yellow) light beams. One of the four versions is called *superimposed display* in which the coloured light beams were superimposed to make a white spot; the spot can be made to appear coloured if one or two of the beams is/are obstructed or controlled by switches. Another version is called *overlapping display* in which the coloured spots were partially separated and made to intersect each other so that colour spots are visible at the periphery and a white spot will be produced at the centre. The *superimposed display* was found to be a spectacular attractive element that generated play, whereas the *overlapping display* made visitors spend longer time on systematic observation. A third version was later created by incorporating features of the two versions and found to be generating playful activities and serious exploration as well. Lucas *et al.* finally conclude that 'we should not focus only on the intended purpose of an exhibit, but need to be alert to unintended exploratory behaviour, which may be as "scientific" as the planned possibilities.'

THE CHANGING MEANING OF PLAY AND LEARNING IN THE INFORMATION AGE

Having examined a number of theories of play and learning, it is necessary now to look at how the meaning of play and learning has undergone changes in the postmodern world. In order to achieve this, we must now turn to what is traditionally considered to be play or learning. Play and learning are often treated as being opposed to each other and they seem to assume different things in relation to concepts of the learner, the learning process, and a number of other topics as shown in Table 7.1.

The nature of play and learning is undergoing drastic changes in the information society, thanks to the marriage of communication technology with computer technology. The shift from an industrial society to an information society created an information-based economy in which emerged a series of computer-oriented people and a range of hitherto unknown jobs. The inclusion of computer games and the concept of 'work at home' facilitated by the information network may well significantly alter the relationship and place of children within the family. Even though information technology, as with any other technological innovation, promises to create more jobs than it abolishes, there will probably be a situation of structural unemployment as a result of its implementation. This is corroborated by the recent emergence of job sharing, shorter working days for staff, longer holidays, sabbaticals, early retirements and a range of others in almost all organizations including museums (Kelly *et al.*, 1994). This means that in the near future there may come a situation when there simply will not be enough work for all. Though this may neither hamper productivity nor the standard of living, it will certainly result in a situation where not everyone may have to work in the way they did

Table 7.1 *Traditional assumptions of play and learning (adapted from Knowles, 1981).*

Topic of Assumption	*Learning*	*Play*
Nature of the process	Explainers- or parents-directed	Free or self-directed
Nature of the person	Dependent personality	Self-directed organism
Readiness	Dictated by agenda or exhibition intents	Develops from tasks and problems
Orientation	Subject- or concept-centred	Task- or problem-centred
Motivation	External rewards and punishments	Internal incentives and curiosity
Cognitive process	Accommodation	Assimilation
Knowledge Domain	Cognitive (mostly)	Affective and psychomotor (mostly)

in the past; rather there would be more choice and more free time. This situation shifts the value of leisure into the centre of people's lives. This shift and the growing awareness of its significance to the quality of life will undoubtedly reinforce the inclusion of leisure activities into educational programmes or educational programmes into leisure activities, which is already happening in museums and in EPCOT-like visitor centres.

The shift in values also represents a range of plural values. The traditional structural 'binaries' change into 'dimensions' in which the dichotomy of binaries fuses and becomes false. In other words, both members of the binary pair are equally valued and considered equally powerful and useful and hence emerge as multiple realities. For example, the recent legal approval of gay and lesbian marriage and their bringing up children fuses the structural binary pair male:female. Many more examples can be given for other binary pairs. Due to the changing nature of leisure and learning in this age, play and learning also seem to fuse. They have become instead different modes through which people engage themselves depending upon their levels of intelligence, interests and experiences. In what follows, a model for the exploration of science in museums is presented by incorporating and integrating all the significant ideas and issues discussed so far.

A 'SPIRAL' MODEL FOR THE EXPLORATION OF SCIENCE IN MUSEUMS

Visitors enter the science museum or science centre with or without their family and during their visit they explore and spend a portion of their two-hour visit in a museum exhibition or an interactive-type exhibition. They are, mostly, on the move, actively exploring to get a sense of the whole museum or centre: they pause casually at most of the exhibits but they give time and pay attention only to a few that they find interesting. The exploration ranges from play to learning, or learning to play, in a cyclical fashion as represented in Figure 7.4.

The tetrahedron in the model represents what the science centre or museum can offer in all three domains of knowledge (science knowledge including). Visitors are expected to climb metaphorically on any or all sides of the tetrahedron. The sides implicitly represent the capabilities and skills each visitor brings with him or

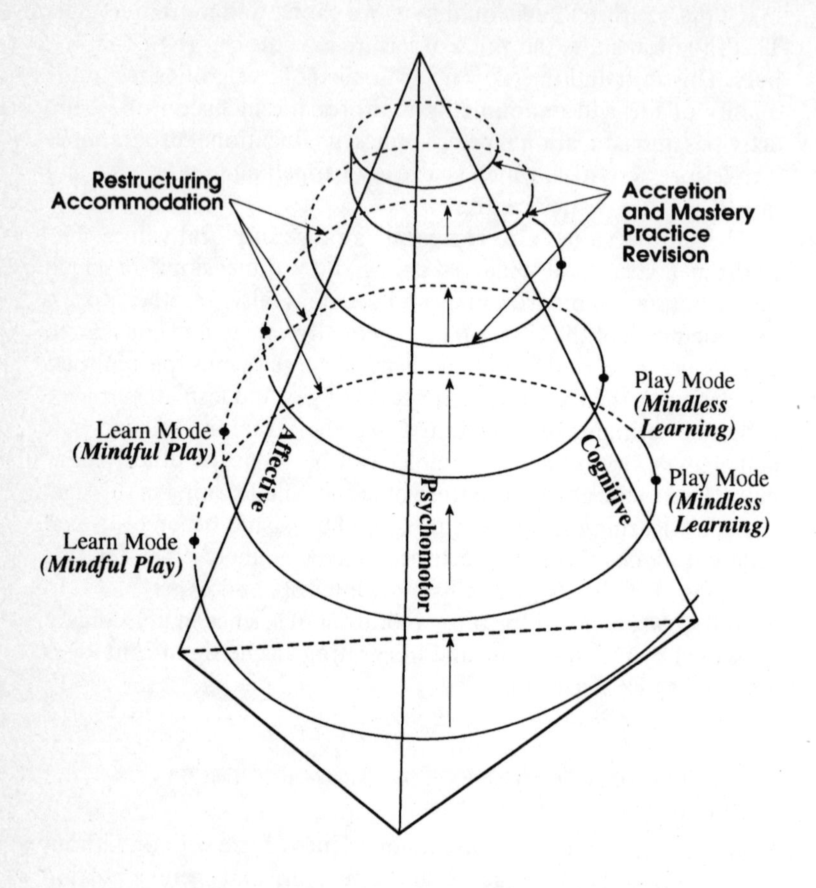

Figure 7.4: A 'spiral' model for the exploration of science in museums. The exploration starts from one of the two modes – *play* and *learn* modes, and switches between them in a spiral fashion embracing the three domains of the knowledge tetrahedron. The *play* mode represents a phase in which the visitor is accretioning or assimilating and mastering skills through practice and problem solving. The *learn* mode represents a phase in which the visitor is restructuring and accommodating the pattern and developing concepts with occasional direction from the companions or facilitators. It is the visitor's choice of modes at will that signifies the uniqueness of the informal learning setting.

her. For example, a sciencc teacher would find it easier to climb up on the side between the vertices 'affective' and 'psychomotor'; an experimental scientist would find it easier to climb up on the side between the vertices 'psychomotor' and 'cognitive'; a scientific philosopher or a theoretical scientist will find it easier to climb up on the side between the vertices 'affective' and 'intellectual'.

The spiral represents how an individual explores science, switching between the two different modes, play and learning, covering all three domains. The slope of the spiral will be more or less steep depending upon the individual and the offerings in each domains. As one goes upwards, the differences between the domains tend to diminish as corroborated by a decrease in playfulness with adults.

This model attempts to accommodate all that goes on inside an interactive-type institution. By arguing that the difference between play and learning is narrow, this model does not entirely reject the traditional concepts of play and learning. It is argued that the structural binary pairs are indeed needed and that they should be treated as classification schemes. At the same time, as Roberts (1992) observed, the traditional meaning of play or learning should not prevent the museum staff from broadening their notions of what constitutes 'education' and 'learning' to include other non-cognitive experiences like social interaction, private reverie and play. Underlying the play and learning modes, as Norman (1978) conceptualized, each visitor undergoes the process of accretion, restructuring and tuning in a simultaneous cyclical fashion. These processes occur at different levels in the visitor depending upon his or her individual capabilities, skills, curiosities, and a range of other styles. Providing a multiple range of opportunities or challenges at different levels in the science centre would facilitate every visitor to explore science in different modes and to match their skills to experience the 'flow'.

CONCLUSIONS

As I set out in the introduction, a spiral approach to exploring science in museums is developed and presented incorporating Piaget's theory of play (*accommodation* and *assimilation*), Norman's theory of complex learning (*accretion, restructuring* and *tuning*), Csikszentmihalyi and Robinson's theory of flow (providing a multiple *range of opportunities and challenges*), Malone's taxonomy of intrinsic motivation (for *curiosity and other factors*), Perry's anatomy of museum visit (for the *cross-section of the knowledge tetrahedron*) and Wellington's three areas of science learning (*cognitive, affective* and *psychomotor* areas). In addition, there are research evidences in the museum field that support this model by arguing that the *interactive* exhibits in science centres do in fact contribute to science learning and that it is possible to provide opportunities for *playful activities* and serious *learning tasks* in a single setting.

Arguing that traditional museum displays make it very difficult for visitors to become actively involved without prior knowledge, Phillips (1986) advocates interactive exhibits that make the visitor (i.e. the explorer) the centre from which all understanding starts and that provide an opportunity for visitors to explore the process of discovery itself. Along the line, Gillies and Wilson (1982) found out that interactive displays are more didactically educational and that they are simply remembered as being more enjoyable than conventional displays. Stevenson (1991), from tracking visitors in the Launch Pad at London Science Museum, found that the visitors do attend to the exhibits for a considerable portion of their time. He also found enough evidence to contradict the view that children spend a large amount of time rushing around. Joan Solomon and Helen Brooke at VISTA, a science centre at Farringdon, England, found that the opportunities for visitors to turn play into learning may be increased by providing a box of equipment beside each exhibit that visitors can use in any way they wanted (Brooke and Solomon, 1992).

The exploration of science in museums can therefore be made more effective by providing opportunities not only to learn but also to play. As Marshall McLuhan (cited in Brill, 1994) put it, 'Anyone who thinks education [learning, a mindful play] and entertainment [play, a mindless learning] are different doesn't know much about either.'

ACKNOWLEDGEMENT

I am thankful to Dr Janet Moyles for it is she who introduced me to the concept of spiral approach to learning and play. I am also grateful to Dr Eilean Hooper-Greenhill for her patient reading of the manuscript and many valuable suggestions.

BIBLIOGRAPHY

AAM (American Association of Museums) (1992) *Excellence and Equity: Education and the Public Dimensions of Museums*, a Report from the American Association of Museums, Washington DC.

Ames, P. (1988) 'A challenge to modern museum management: reconciling mission and market', *Museum Studies Journal*, Spring–Summer.

Annis, S. (1986) 'The museum as a staging ground of symbolic action', *Museum*, 38(3): 168–71.

Bagchi, S.K., Yahya, I., and Cole, P.R. (1992) 'Piagetian children's science gallery', *Curator*, June.

Baird, D.M. (1986) 'Science museums in the modern world', *Curator*, 29(3): 213–20.

Brill, L.M. (1994) 'Museum-VR: opening the gateway to cyberspace. Part I', *Virtual Reality World*, November/December: 33.

Brooke, H. and Solomon, J. (1992) 'Play or learning? how can primary pupils benefit from an interactive science centre?', *EiS*, January: 16–17.

Carnes, A. (1986) 'Showplace, playground or forum? Choice point for science museums', *Museum News*, 64(4): 29–35.

Cohen, D. (1993) *The Development of Play*, 2nd edn (London: Routledge).

Csikszentmihalyi, M. and Robinson, R.E. (1975) *Beyond Boredom and Anxiety* (San Francisco: Jossey-Bass).

Danilov, V.J. (1976) 'America's contemporary science museums', *Museums Journal*, 76: 75.

Danilov, V. (1989) 'The Exploratorium of San Francisco twenty years later', *Museum* (UNESCO), p. 163.

Falk, J.H. and Dierking, L.D. (1992) *The Museum Experience* (Washington DC: Whalesback).

Gardner, H. (1991) *The Unschooled Mind: How Children Think and How Schools should Teach* (California: J. Paul Getty Museum).

Gillies, P.A. and Willson, A.W. (1982) 'Participatory exhibits: is fun educational?', *Museum Journal*, 82(3): 131–4.

Gilmore, J.A. (1985) 'Play: a special behaviour' in R.E. Herron and B. Sutton-Smith (eds) *Child's Play* (Malabar, FL: Robert E. Kreiger).

Golley, F. (1988) 'Great ideas in science – the ecosystem' in P. Heltne and L. Marquardt (eds) *Science Learning in the Informal Setting* (Chicago, IL: Chicago Academy of Sciences).

Hooper-Greenhill, E. (1994) *Museums and their Visitors* London: Routledge.

Hudson, K. (1987) *Museums of Influence: Science, Technology and Industry* (Cambridge: Cambridge University Press).

Huizinga, J. (1951) *Homo Ludens* (Boston, MA: Beacon).

Johnson, C. (1992) 'Hands-on science centres and teacher education', *Science Teacher Education*, 6 (September).

Kelly, J. *et al.* (1994) 'New work patterns', *Museums Journal*, October: 23–6.

Kimche, L. (1978) 'Science centres: a potential for learning', *Science*, 199: 270–3.

Knowles, M. (1981) 'Andragogy' in Collins, Z.W. (ed.) *Museums, Adults and the Humanities: A Guide for Educational Programming* (Washington DC: American Association of Museums), pp. 49–60.

Kubie, L.S. (1971) *Neurotic Distortion of the Creative Process* (New York: Noonday Press).

Lucas, A.M., McManus, P., and Thomas, G. (1986) 'Investigating learning from informal sources: listening to conversations and observing play in science museums', *European Journal of Science Education*, 8(4): 341–52.

MacDonald, G. (1988) 'Epcot centre in museological perspective', *Muse*, Spring.

Malone, T.W. (1980) *What Makes Things Fun to Learn? A Study of Intrinsically Motivating Computer Games*, Monograph no. CIS-7 (SSL-80–11) (Palo Alto, CA: Xerox Research Centre).

Malone, T.W. and Lepper, M.R. (1987) 'Making learning fun: a taxonomy of intrinsic motivations for learning' in R.E. Snow and M.J. Farr (eds) *Aptitude, Learning, and Instruction, vol. 3: Conative and affective process analyses* (Hillslade, NJ: Lawrence Erlbaum) pp. 223–53.

McManus, P.M. (1989) 'Oh yes they do! How visitors read labels and interact with exhibit texts', *Curator*, 32(3): 174–89.

Moyles, J. (1989) *Just Playing? The Role and Status of Play in Early Childhood Education* (Milton Keynes and Philadelphia: Open University Press).

Norman, D.A. (1978) 'Notes towards a complex theory of learning' in A.M. Lesgood (ed.) *Cognitive Psychology and Instructions* (New York: Plenum).

Orchistron, W. and Bhathal, R.S. (1984) 'Introducing the science centrum: a new type of science museum', *Curator*, 27(1): 33–47.

Parkyn, M. (1993) 'Scientific imaging', *Museums Journal*, 93(10): 29–34.

Pearce, S. (1989) 'Objects in structures' in S. Pearce (ed.) *Museum Studies in Material Culture* (Leicester and London: Leicester University Press), pp. 47–60.

Pearce, S. (1992) *Museums, Objects and Collections: A Cultural Study* (Leicester and London: Leicester University Press).

Perry, D.L. (1989) 'The creation and verification of a development model for the design of a museum exhibit,' unpublished PhD thesis, Indiana University.

Perry, D.L. (1992) 'Designing exhibits that motivate', *ASTC Newsletter*, March/April: 9–12.

Perry, D.L. (1993) 'Beyond cognition and affect' in D. Thompson, S. Bitgood, A. Benefield, H.H. Shettel and R. Williams (eds) *Visitor Studies: Theory, Research and Practice, vol. 6* (Jacksonville, AL: Visitor Studies Association), pp. 43–7.

Phillips, D. (1986) 'Science centres: a lesson for art galleries?' *International Journal of Museum Management and Curatorship*, 5: 259–66.

Piaget, J. (1951) *Play, Dreams and Imitation in Childhood* (London: Heinemann).

Porter, G. (1993) 'Alternative perspectives', *Museums Journal*, 93: 11 25–7.

Quin, M. (1991) 'All grown up and ready to play', *New Scientist*, 26 October: 60–1.

Quin, M. (1994) 'Aims and strengths and weaknesses of the European science centre movement' in R.S. Miles and L. Zavala (eds) *Towards Museum of the Future: New European Perspectives* (London: Routledge).

Roberts, L. (1992) 'Affective learning, affective experience: what does it have to do with museum education?' in A. Benefield, S. Bitgood and H. Shettel (eds) *Visitor Studies: Theory, Research, and Practice, vol. 4* (Jacksonville, AL: Center for Social Design).

Samples, B. (1976) *The Metaphoric Mind: A Celebration of Creative Consciousness*. (Reading, MA: Addison-Wesley).

Shortland, M. (1987) 'No business like show business', *Nature*, 328: 213–14.

Stevenson, J. (1991) 'The long term impact of interactive exhibits', *International Journal of Science Education*, 13(5): 521–31.

Sutton-Smith, B. (1971) 'A reply to Piaget: a play theory of copy' in R.E. Herron and B. Sutton-Smith (eds) *Child's Play* (Malabar, FL: Robert E. Kreiger).

Tressel, G.W. (1980) 'The role of museums in science education', *Science Education*, 64(2): 257–60.

Wellington, J. (1989) 'Attitudes before understanding: the contribution of interactive centres to science education' in M. Quin (ed.) *Sharing Science: Issues in the Development of Interactive Science and Technology Centres* (London: Nuffield Foundation and COPUS).

Wellington, J. (1990) 'It's fun, but do they learn?' *Questions*, February.

Wymer, P. (1991) 'Never mind the science, feel the experience', *New Scientist*, 5 October: 53.

Yahya, I. (1989) 'Objecta: none unique nor aesthetic object of science museums' in H. Bedekar (ed.) *New Museology* (New Delhi: Museum Association of India).

Yahya, I. (1994) 'Snibston Discovery Park: a review from an Indian perspective' in S. Pearce (ed.) *New Research in Museum Studies, An International Series, no. 4: Museums and the Appropriation of Culture* (London: Athlone Press).

8

Science Museums, or Just Museums of Science?

JOHN DURANT

INTRODUCTION

Over the past few years, museums have been changing rather rapidly. So far as museums of science are concerned – and here I include both specialist science and technology museums and general museums that have significant scientific and technological interests – the incentive both to do differently and to do better has come from two different but related quarters. From the visitor attraction industry itself, the obvious success of the world-wide 'science centre movement' has served to raise doubts about the very need for science museums, by demonstrating that it is perfectly possible to attract large numbers of visitors to interactive, 'hands on' exhibits that are entirely independent of costly and cumbersome museum collections; and from the scientific and the educational communities, the world-wide 'public understanding of science movement' has served to increase the pressure on museums of science to contribute more directly to the creation of a scientifically literate society (Schiele, Amyot and Benoit, 1994).

If museums of science are to move forward, they will need to be clear-headed about their own strengths and weaknesses, and far-sighted about their place in society. This essay assesses the problems and prospects of museums of science in the context of recent research of several different kinds. It starts by identifying a number of different traditions of research in or about the interpretation of science in museums; it goes on to consider the particular nature of the challenge that is posed to museums by science; it continues with an assessment of different strategies of scientific interpretation in museums; and it concludes by offering a vision of what museums of science could and should become in order to fulfil their true potential. It should be admitted at the outset that this vision is personal, and probably not to everyone's taste. I make no apology, however, for arguing that museums of science must become much more than mere 'museums of science' if they are to enjoy a secure place in the future. The museum of science of tomorrow will be a very different sort of place from the museum of science of today; and its differentness will reflect in part a re-definition of the nature of the museum itself.

THE CHALLENGE OF SCIENCE

It is one thing to explore science in museums as a visitor, but quite another to do so as a museologist. The museum visitor engages in a process of exploration that is largely informal and self- (or, more likely, group-) directed; but the museologist is under an obligation to proceed more formally, paying due deference to both the knowledge base and the professional standards of museum studies. In the case of science, at least, the museum studies knowledge base is relatively slight (of this, more later); but so far as professional standards are concerned, a choice has to be made. For to some, museum studies is a branch of cultural studies governed by the (somewhat nebulous, but broadly speaking literary) rules of cultural criticism; whereas to others, museum studies is a branch of social study (commonly known as 'visitor studies') governed by its own rather particular, albeit fairly eclectic, rules of investigation.

Between these two conceptions of museum studies there is often precious little mutual sympathy or respect. To the advocates of cultural studies, visitor studies are apt to seem crassly empiricist and narrowly positivist; while to the advocates of visitor studies, cultural studies tend to appear ill-disciplined, unreliable and even self-indulgent. It does not help, of course, that the contrast between these two conceptions mirrors in the microcosm of museum studies the much larger tension that exists within the macrocosm of sociology (or 'social science': the very name of the discipline is itself contentious). For many years, Anglo-Saxon sociology has been divided into broadly humanities-oriented (qualitative) and broadly science-oriented (quantitative) traditions of enquiry. Unlike its continental counterparts, which appear rather more comfortable with a plurality of theoretical and methodological resources, home-grown sociology commonly presents itself as so many alternative camps or schools of thought, so that to be trained as a sociologist in one institution rather than another is often to be inducted into the ways of one discipline rather than another.

The nature of the divide within museum studies is well illustrated by two contrasting contributions to a recent essay collection, *Visitor Studies in the 1990s* (Bicknell and Farmelo, 1993). In an abridged version of an article that had already appeared in an earlier volume of museum studies, one of my colleagues in the Science Museum, Senior Curator of Clinical Medicine Ghislaine Lawrence, set out to assess 'the appropriateness of the "traditional scientific arena" with its underlying positivist assumptions to the study of cultural and social processes'. Criticizing the museological tradition of exhibit evaluation for being not only 'positivist' but also 'empiricist' and 'behaviourist', Lawrence recommended instead 'those methods used to analyse cultural representations where connotational meaning is central – the methods of art, literary, film or historical criticism' (Lawrence, 1993). In reply, the Natural History Museum's Roger Miles criticized Lawrence's 'behaviourist parody of museum evaluation', questioned the credibility of her proposed alternatives to the 'traditional scientific arena', and reasserted the value of the disciplined and objective assessment of exhibit effectiveness (Miles, 1993).

The relevance of this debate to the theme of the present volume is clear. We have to deal, it would seem, not merely with rival museologies but also with rival conceptions of the nature of science itself. For Lawrence, science is problematical: its methods and theories are historically constrained; and those who plunder it for intellectual resources with which to conduct museum studies are likely to fall victim to its inherent limitations. For Miles, by contrast, science is merely the term we use for rational exploration: scientific methods and theories are to be judged entirely on the basis of their suitability to the purposes at hand ('origins are irrelevant'); and to be scientific is merely to be willing to engage in open-ended exploration using the best available method(s) of investigation. In short, Lawrence's approach to science itself is critical, while Miles's is pragmatic; Lawrence invites us to engage in historical and sociological reflection upon the nature of the interpretive processes that are necessarily involved in exhibiting science, while Miles encourages us to be both visitor-centred and disciplined in our assessment of the effectiveness of our interpretive efforts.

It is not at all clear to me why we must choose between these radically different alternatives. Disciplined assessment of the extent to which aims and ambitions are realized in particular exhibits is obviously a good thing; but so too is critical reflection on the general character of exhibits. Miles helps us to see a way forward by distinguishing carefully between *evaluation* and *research* in museums. Evaluation, he suggests, aims to make practical judgements on a largely pragmatic basis; whereas research aims to make generalizations on the basis of rigorous enquiry. Clearly, there is overlap between these two activities: generalizations derived from research may be of great assistance in evaluation; and more ambitious evaluation, performed for purely practical purposes, may aspire to the discovery of useful generalizations. Nevertheless, the contrast between two related but distinct activities seems helpful, and to it I would add a third: namely, *criticism.* Given Miles's seemingly unobjectionable definitions of evaluation and research, it is far from clear that Lawrence is advocating either. Instead, her reference to 'the methods of art, literary, film or historical criticism' invokes a tradition that is concerned more with the exercise of critical judgement in individual cases than it is with the establishment of general principles or laws that apply to entire classes. Once again, I see no need to question the value of this particular tradition; but it is clearly rather different from the other two.

The threefold distinction between evaluation, research and criticism serves to capture most of what is done under the banner of research and scholarship in the field of museum studies. Evaluation covers all of the work with visitors and exhibits that is used directly to inform the processes of exhibition and gallery development. Research covers a rather smaller number of generally more ambitious and systematic studies that are intended to resolve – or at any rate to raise – fundamental questions concerning the nature or the role of museums; and criticism covers most of the more or less well-informed commentary that inevitably attaches to museum productions of all kinds. Of the three categories, research is the only one that is arguably the exclusive preserve of academically trained specialists. Evaluation is undertaken both by specialist evaluators and by a fairly wide range of other museum professionals; and similarly, criticism is undertaken both by academic commentators and by journalists and an assortment of other museum camp-followers. Notoriously, everyone feels entitled to criticize.

My purpose in characterizing the field of museum studies in this way is twofold: first, I want to enter a plea for a truce in the largely sterile conflict between evaluators and critics that has gone on for many years within museum studies; and second, I want to point to the inevitably reflexive nature of museum studies devoted to the interpretation of science. By reflexivity in this context, I mean simply that museum studies of science involve the conduct of scientific investigation into the representation of scientific investigation in museum displays. Such activity cannot avoid incorporating assumptions about the object under study into the study itself. Whether or not evaluators and critics tend to prefer different types of scientific museology, evaluation and criticism certainly embody different conceptions of the scientific enterprise. Scientific museology can only benefit from a greater professional awareness of the choices that are explicitly or implicitly made in the selection of one approach over another in this field. The nature of some of these choices is made clearer by a consideration of the distinctive problems that are faced by museums of science.

THE CHALLENGE OF EXPLORING SCIENCE IN MUSEUMS

By its very nature, science poses extraordinary challenges to museums. These challenges are so daunting that one might be forgiven for wondering why anybody would wish to establish a museum of science at all. In truth, not many have done so. The number of great national museums that specialize in science is extremely small (arguably, there are only a handful in western Europe and none at all in North America); and of these, the greater number – including my own – were established as much from a concern for technology and industry as from any particular love of science *per se*. In fact, the only museums of which I am aware that deal mainly or wholly with science are small, university-based institutions such as the Museum of the History of Science in Oxford and the Whipple Museum in Cambridge.

Of course, there is nothing wrong with linking science, technology and industry. On the contrary, such linkage reflects one of the important realities of a modern world in which the pursuit of theoretical knowledge and the acquisition of practical power have become closely intertwined. Nonetheless, the dominance of technology in many so-called science museums is extremely significant.

It testifies not only to the interdependence of science and technology in the modern world but also to the fact that in many ways technology appears more suited than science to the museological arts of curatorship and interpretation. By its concreteness, by its visual impressiveness, by its obvious usefulness; above all, perhaps, by its frequently heroic mastery of the forces of nature, technology lends itself to the purpose of display. Why else, after all, are the pyramids undisputed wonders of the ancient world, if not because they are the technological marvels of their age? So long as people wish to see the finest fruits of human inventiveness preserved and celebrated, so long will there be a demand for museums of technology and industry.

But what of science? By contrast with technology, science is abstract rather than concrete, unobtrusive rather than visually impressive, and apparently pointless rather than obviously useful. Where technology aims to master the forces of nature, science seeks merely to delineate them. In this sense, science is exciting only to those whose pulses quicken at the prospect of intellectual discovery. Such discovery is the fruit of considerable rigour, and typically it takes the form of general propositions, often mathematical in form, that may be inaccessible to all but the most thoroughly initiated. Many of Isaac Newton's contemporaries found his *Principia Mathematica* dauntingly difficult; and as if the laws of motion and universal gravitation were not bad enough, Albert Einstein showed that these were mere first approximations to more fundamental laws of space and time which most people today – even, I dare say, most university graduates – find dauntingly enigmatic.

The sheer intellectual difficulty of science is one of the factors that is often cited in lists of the reasons why interpreting science in museums is especially challenging. Other commonly cited reasons are that science deals for the most part with the non-human rather than the human world, that scientists tend to employ their own highly specialist vocabularies, that the pace of scientific change is bewilderingly fast, and even that science is (or at any rate has been) a largely male-dominated activity. I have no particular wish to deny any of these reasons here – though it is perhaps worth noting in passing that male domination is hardly confined to science – but I do want to suggest that there are a number of other, less often noticed difficulties that face those of us who work in science museums. These difficulties flow from inherent tensions between the nature of science as a process of discovery and the nature of the museum as a repository of material culture.

Science is both an intellectual and a practical activity, but scientists tend to set little store by their own material culture – by the places and the paraphernalia, if you will, of their professional lives. This is partly because the true object of attention in science is knowledge. With some exceptions – for example, microscopes and telescopes – scientific and technical apparatus has little residual value once it ceases to be of immediate use in the knowledge industry. This means that museums of science can often acquire objects at little or no cost – indeed, a key curatorial task consists in saving potentially important objects from the scrap-heap; but it also means that museums of science face the challenge of securing – as opposed to merely celebrating – the intrinsic value of their collections. Paradoxically, this problem appears greater the closer we come to the present – an abacus and an astrolabe, for example, have more obvious museological attractions than a diode or a DNA sequencer.

Related to the problem of science's relative indifference to its own material culture is the problem of its relative lack of interest in its own history. The history of architecture and of art are extremely well-established areas of study, not least because practising architects and artists generally recognize them as essential sources of insight. By contrast, the history of science is far less securely placed in the world of science; and this is because practising scientists rarely acknowledge the relevance of history to their work. Generally speaking, what science regards itself as inheriting from the past is merely a limited stock of (historically unproblematic) knowledge, together with an array of questions requiring further exploration. Yesterday's achievements are of little inherent interest; they merely set the stage for what is to come. This sort of scientific culture is not easy to reconcile with the traditional culture of the museum, in which past accomplishments rather than present activities or future prospects have generally held centre stage.

There is a final problem that confronts all who work in museums of science, at least in the Anglo-Saxon world, and this is that while museums have generally been taken to be concerned with the collection and display of various aspects of culture, science has not been widely recognized as being a part of culture at all. Here, much depends upon the meaning we attach to the term 'culture'. Rather than enter into matters of definition at this point, I simply observe that for at least a century there has been debate about the apparent isolation of science from other areas of learning – and, not least, from the arts, the classics and the humanities. This debate has been particularly prominent in Britain, extending from mid-nineteenth-century arguments between Matthew Arnold and Thomas Huxley about the proper nature and scope of education, through mid-twentieth-century exchanges between F.R. Leavis and C.P. Snow about 'the two cultures', to current debates about the allegedly 'anti-scientific' cast of certain academic and journalistic commentaries on science (see, for example, Wolpert, 1993). While the balance of power between the two cultures may have shifted considerably over the past half century, there is no evidence that today they are any closer to reunification.

Of course, the isolation of science from the wider culture – its separation from history, from the arts and humanities, and even from politics – is not a problem peculiar to museums; but museums are peculiarly affected by it. Traditionally, science has been something to respect, to defer to, to strive to understand; it has not been something to admire, to participate in, or even to criticize. In these senses, science has tended to be less inviting than most other museological subjects. In contrast to the arts, science has often been portrayed as something that one either takes or leaves, rather than something to which one is invited to make some sort of personal response. This, I believe, accounts very largely for the sense of personal failure that some commentators have observed to be the common result of a visit to a museum of science (see, for example, Kavanagh, 1992). If the museum visitor accepts the role of passive recipient (as opposed, say, to that of active participant or critic), then he or she is likely to be overwhelmed by the dead weight of scientific authority. The remedy for this is not (as some have supposed) cleverer use of graphics or multimedia, but rather a change in the visitor's perception of his or her role in relation to science; and such a change presupposes the possibility of thinking about science in a different way.

CONTRASTING STRATEGIES OF SCIENTIFIC INTERPRETATION

In the past, museums have responded to the challenges posed by science in a rather characteristic way. Concentrating for the most part on the history of a number of key technologies (e.g. energy production and conversion, transport, communication), they have offered a largely celebratory account of humankind's progressive mastery of the natural world. Such an account, I suggest, informs the bulk of the work undertaken by institutions such as the Centre National des Arts et Metiers in Paris, the Deutsches Museum in Munich and the Science Museum in London, since their foundation in the eighteenth and nineteenth centuries. Relatively distanced from the practice of science itself, these institutions have offered visitors a particular historical vision of the nature and development of modern industrial society.

Advocates of this traditional approach have pointed to its extensive use of museum collections, as well as to its evident popularity in areas (such as transport) where such collections lend themselves to bold and striking display. Critics, on the other hand, have pointed to the largely non-interactive nature of traditional galleries, as well as to the dependence of such galleries on highly questionable historiographical assumptions concerning the nature and consequences of scientific and technological development. Interestingly, two very different research traditions – the one educational, the other historical – have supported the critics' case. Educationalists have long argued that concrete experience is a more effective route to learning than abstract contemplation, and historians have long since abandoned the charting of linear progress in science and technology in favour of contextual studies of science in society. Where the educationalists' preoccupation with experience has inspired the so-called 'hands-on' science movement (see, for example, Gregory, 1989), the historians' preoccupation with context has led to a general loss of confidence in traditional 'Plato to NATO' treatments, but no clear consensus about what to do instead has yet emerged. It is worth saying a little more about each of these distinct responses to the traditional approach.

The 'hands-on' science movement has been – and remains – the single most potent force for change in museums of science. The movement may be traced back as far as the 1930s, when the Children's Gallery at the Science Museum, London, introduced the world to push-button working models; but more immediately, it has its roots in North America in the 1960s, and especially in the innovative work of Frank Oppenheimer at the Exploratorium in San Francisco. Oppenheimer's aim was to enable people to feel 'at home' with scientific insights into the natural world, by giving them the opportunity to develop these insights for themselves through first-hand experience (Oppenheimer, 1973). This basic philosophy has inspired an entire generation of exhibit constructors and science centre developers. Its basic appeal to visitors is immediately evident in any museum of science that combines traditional with hands-on displays – for the most part, the traditional displays are relatively empty, whereas the hands-on displays are relatively packed with visitors. 'Launch Pad' and 'Flight Lab,' the two areas of dedicated 'hands on' interactives in the Science Museum, London, at the time of writing (three more such areas are due to open in the autumn of 1995) are the only areas of the entire museum in which time-ticketing has to be employed in order to cope with visitor demand.

The 'hands-on' approach has generally discarded both museum collections and history in its preoccupation with immediate sensory experience. For this reason, most genuine museums of science have combined the adoption of a certain amount of 'hands on' exhibitry with the quest for new ways of interpreting their historical collections. If any trend is evident in this area, then perhaps it is a tendency towards the conventions of social history. Consider, for example, the new and much-discussed exhibition at the Smithsonian Institution's National Museum of American History in Washington DC on 'Science in American Life' (Lewenstein, forthcoming). This exhibition has attracted considerable criticism from the American scientific community, not least because it deals with both the technical and the social dimensions of major scientific developments such as atomic physics and human genetics. Where a traditional treatment might have presented the transition from early-twentieth century studies of atomic structure to the development of military and civil nuclear power as a relatively unproblematic progression from knowledge to power, the social history perspective raises difficult moral and political questions concerning the wider impact of science and technology upon society. That some scientists have expressed deep misgivings about 'Science in American Life' illustrates the potential for conflict that is inherent in the museological decision to interpret the social significance of science and technology in any but strictly scientific and technological terms.

BEYOND MUSEUMS OF SCIENCE

Museums are the products of particular cultures. Most museums of science were founded by industrial cultures that were imbued with a sense of great optimism about science and technology. Today, that sense of optimism still exists (just think, for example, of the way in which new developments in the biomedical sciences are reported in the mass media); but it is tempered by a growing awareness of the perils and pitfalls of knowledge. Ours is an age of science; but it is also an age of the critical reappraisal of science. The impetus for reappraisal comes from several directions. First, there is the sheer success of the scientific enterprise itself, which has grown under the Baconian maxim that 'knowledge is power' to the point at which it invites the sort of public scrutiny that inevitably attaches to other powerful social institutions. Second, there is the fact that many important areas of modern science and technology (e.g. information technology, and biotechnology) clearly raise difficult moral, social and political issues; issues, moreover, which scientific expertise alone is insufficient to resolve. Third, there is the growing recognition that scientific knowledge claims are frequently matters of doubt and uncertainty, particularly in cases where such claims bear directly on matters of public policy.

In cultures that regard science and technology as unproblematic social goods, there is perhaps a role for museums of science as chroniclers of change. In a culture such as ours, however, in which the social contract between science and society is under review, this role requires radical reassessment. Today, any museological attempt simply to plot the path of scientific and technological progress would fail to do justice to the nature of public interest in and concern about the place of science and technology in society. The public agenda for science and technology – as evidenced, for example, by contemporary debates on information technology, on recombinant DNA technology, and on the global environment – extends well beyond the professional boundaries of the scientific community; indeed, it extends well beyond all professional boundaries, to embrace fundamental questions about the relationship between 'experts' and 'lay people' within democratic cultures. If they are to live up to their responsibilities and fulfil their true potential, science museums must find new ways of exploiting their distinctive strengths in order to address the changing public agenda for science and technology.

There can be no single formula for this sort of development. However, I should like to offer a number of candidate general principles which seem to me to be relevant. First, science museums must find ways of maintaining an appropriate balance between past and present in their galleries and public programmes. The past is important, not least because it provides alternative perspectives on the present; but the present deserves a prominent place of its own. The high capital costs and long lead times of so-called 'permanent galleries' make them difficult vehicles for fast-moving areas of science and technology. However, updating can (and should) be built into even the largest gallery developments; and alongside such developments, science museums should introduce smaller, cheaper exhibitions which can be produced in months rather than years. This is the rationale of the Science Museum's 'Science Box' series of contemporary science exhibitions. A flexible, re-usable exhibition system allows the museum to mount three new exhibitions a year on topical subjects such as upper atmosphere ozone depletion, the development of new treatments for infertility and the prospects for the Information Superhighway.

A second general principle is that science museums should find as many different ways as possible of involving visitors in science and technology. Involvement can take many forms. Interactivity is the obvious and most commonly used technique here: simple interactives allow individual visitors to engage with exhibits; but more complex ones can encourage visitors to engage with one another in co-operative (or competitive) exploratory games. In addition to interactive exhibits, gallery programmes ('Explainer' support for interactives, demonstrations, workshops, plays, tours, evening and overnight events, etc.) all help to make museums places where people come to do and not merely to see things. Obviously, space must be provided for visitors who prefer to spend part or all of their time in quieter, more contemplative activities; but it is striking how many visitors to science museums gravitate to places where they can be players rather than mere spectators.

A third general principle is that science museums should think very carefully about the 'tone of voice' in which they communicate with their visitors. In the past, the tone was variously scholarly (i.e. austere), schoolmasterly (i.e. arid) and sensational (i.e. accessible). In each case, the tone of voice placed museums in positions of almost magisterial authority – they knew, and the visitors' task was to learn. Certainly, visitors need to be able to trust science museums as competent and credible sources of information; but they do not need to go away with the impression that science museums are founts of all wisdom on matters scientific. For most purposes, I would judge that the authoritarian tone of voice is ill-suited to a contemporary culture which is accustomed to hearing (and questioning) many different points of view, and which is much less inclined to defer to acknowledged experts than may have been the case in the past. There are many credible alternatives to the authoritarian tone of voice: the questioner; the evaluator; the critic; etc. All of these are worth experimenting with in science museums.

A fourth and final general principle that I would offer is that science museums should be a good deal braver in their choice both of exhibition topics and of interpretive techniques. Paradoxically, genuine experimentation in the art of communication with visitors is more common in museums of art than it is in museums of science. In their effort to challenge and provoke visitors, exhibitors of art commonly resort to subjects and styles that are intended to be amusing, disturbing or even shocking. By contrast, exhibitors of science appear generally far more conservative. Weighed down by perceived obligations to be accurate, clear and comprehensive, they commonly stick to the relatively safe ground of well-established topics and well-tried interpretive techniques. A period of conscious risk-taking in exhibition development is long overdue in science museums.

CONCLUSION

Science museums are uniquely well placed to serve a public which is both fascinated and slightly frightened by science and technology. Their collections provide an unparalleled resource for interpreting a notoriously inaccessible aspect of modern industrial culture; and their exhibitions and programmes offer an invaluable point of entry into the world of science for visitors who feel remote and even alienated from it. Science museums have the potential to be a meeting ground between the scientific community and the public; but if they are to realize this potential, they must break free from older ways of working that were suited to more deferential and less questioning times. In short, the challenge is to turn mere museums of science into genuine science museums.

BIBLIOGRAPHY

Bicknell, S. and Farmelo, G. (eds) (1993) *Museum Visitor Studies in the 90s* (London: Science Museum).

Durant, J. (ed.) (1992) *Museums and the Public Understanding of Science* (London: Science Museum in association with the Committee on the Public Understanding of Science).

Gregory, R. (1989) 'Turning minds on to science by hands-on exploration: the natural and potential of the hands-on medium' in M. Quinn (ed.) *Sharing Science. Issues in the Development of Interactive Science and Technology Centres* (London: Nuffield Foundation in association with the Committee on the Public Understanding of Science), pp. 1–9.

Kavanagh, G. (1992) 'Dreams and nightmares: science museum provision in Britain', in J. Durant (ed.) *Museums and the Public Understanding of Science* (London: Science Museum), pp. 81–7.

Lawrence, G. (1993) 'Remembering rats, considering culture: perspectives on museum evaluation' in S. Bicknell and G. Farmelo (eds) *Museum Visitor Studies in the 90s* (London: Science Museum), pp. 117–24.

Lewenstein, B. (forthcoming) 'Is there anything wrong with Science in American Life?', *Science Communication.*

Miles, R. (1993) 'Grasping the greased pig: evaluation of educational exhibits' in S. Bicknell and G. Farmelo (eds) *Museum Visitor Studies in the 90s* (London: Science Museum), pp. 24–33.

Oppenheimer, F. (1973) 'Everyone is you . . . or me', M. Quinn (ed.) *Sharing Science: Issues in the Development of Interactive Science and Technology Centres* (London: Nuffield Foundation in Association with the Committee on the Public Understanding of Science) p. 1.

Schiele, B., Amyot, M. and Benoit, C. (eds) (1994) *When Science Becomes Culture: Wold Survey of Scientific Culture* (Quebec: University of Ottawa Press).

Wolpert, L. (1993) *The Unnatural Nature of Science* (London and Boston: Faber).

PART TWO

Reviews edited by
Eilean Hooper–Greenhill

Australian Interventions: 'The Artist and the Museum' Series, Ian Potter Gallery, University of Melbourne

REBECCA DUCLOS
Museum Studies Programme, University of Toronto

THE 'ARTIST AND THE MUSEUM' SERIES

The Artist and the Museum series was proposed by Merryn Gates, Assistant Director at the Ian Potter Gallery. Gates provided the curatorial rationale as well as the spatial forum in which artists Aleks Danko, Elizabeth Gertsakis, and Black Day Dawning (collective) could offer personal responses to the complexity of issues which occupied them as visual artists, cultural theorists and Australian citizens of diverse heritage. With this series the curator encouraged artists to link their personal explorations with institutional collections. The resulting investigations had a refreshingly complex character: private issues of cultural identity, political ideology and artistic practice were 'worked through', using – and sometimes subverting – the public culture of the museum and the gallery.

For museologists, interventions such as these have obvious appeal. Not only do alternative approaches allow the abstract theoretical concepts of our discipline to be innovatively and concretely realized, they introduce traditional practices of curatorship to a whole new world of interpretive possibilities. While we are quite familiar with the talk of much museum studies literature – our institutions operate through practices of inclusion and exclusion; they are sites for the construction and dissemination of knowledge; objects are the territories upon which the semiotic interplay of sign and symbol takes place – we are perhaps less sure of how to turn our critical speech into curatorial practice. As the museum community becomes more involved in interdisciplinary explorations of thought and intrigued by interpretive productions in diverse media, institutional collections are having the dust creatively blown off them. At the heart of some of this activity are visual, conceptual and performance artists who have either been invited into collections or have been independently inspired by their work with objectival and archival evidence.

Danko, Gertsakis, King-Smith, Black and Huggins were perhaps obvious choices for The Artist in the Museum series because their work very obviously places interpretive and ideological issues at the fore. A decade of activity by Danko reveals the artist's success at creating playfully subversive critiques of modernist (institutional) ideals. Gertsakis, a Greek Australian, uses the museum as a place of research and the gallery walls as the site for her reframed investigations into 'national identity'. The Black Day Dawning collective, set up by an Aboriginal and Torres Strait Islander women's art organization, encourages women to use the arts as an alternative way to tell their stories and reconnect with their objects.

THE THREE TROPES: CRISIS. MUSEUMS. TREACHERY.

The Artist and the Museum series, in allowing for a variety of viewpoints and diversity of approaches to be expressed within a collective programme, represents the first cohesive attempt in Australia to explore and expand the parameters of museologically based artwork. Supplementing and further qualifying the aesthetic presence of the individual installations, the critical commentary which accompanies the series is so pertinent it might well become required reading for museum studies students. Along with the two excellent essays in the *Projects For Two Museums* catalogue (Barrett-Lennard, 1993; Barnes, 1993), Charles Green's contribution to *Zen Made In Australia* (Green, 1994) is notable. In his analysis of themes in Danko's work, Green's introduction is bold enough to name three elements which characterize the genre of institutional critique. 'Crisis' 'Museums' and 'Treachery'. What a list! And yet his three 'tropes' work well to gather up threads which are woven into each of the series works.

The pieces created by Danko, Gertsakis and Black Day Dawning in some way touch upon the 'crises' of contemporary society.

Crises of identity, of loss and exclusion, misrepresentation and non-communication, fragmentation and repression are addressed in each of the works. The location where these crises are confronted and to some extent resolved is within the museum – or more correctly – within the museum's spaces that are pried open for personal and artistic intervention. The impetus for confrontation and the mode of redress is quite often, and without apology, treacherous.

Crisis

What is the crisis to which Danko, Gertsakis and Black Day Dawning are responding? Or are they, by virtue of their interventionist, manipulative stance, actually creating a crisis for their audiences to confront?

Elizabeth Gertsakis' installation, 'Beyond Missolonghi,' incorporates historical paintings, pencil sketches, colour bubble jet prints and vinyl letter texts in an effort to bring together the incongruity of sources, images and language metaphors that have helped to construct fictions of the Ottoman world, Greece, Macedonia, the Balkans and Islam. In juxtaposing pieces created by artists in the nineteenth century with those assembled contemporarily in her own studio, Gertsakis forces the viewer to make the link between images and ideologies, between heroic depictions and virulent nationalism. In this artist's hands, the visual image- metaphors unearthed in national collections at the Galleries of New South Wales, South Australia and Victoria are re-exhibited to describe how images and fantasies of 'the other' are used to promote versions of one's own identity.

The 'crisis' which Gertsakis invokes is indeed one of identity, but on a much larger scale she is also concerned with the ideology of representation. This is a political-cultural territory which museums, by virtue of their interpretive stance, have enormous power to manoeuvre within. Gertsakis challenges them to do so in a much more provocative and searching way. 'Beyond Missolonghi', in the words of Nikos Papastergiadis,

> begins by bringing up what was held in the unconscious of the Museum. Entering the first room is like bearing witness to the bizarre dispersal of images that formed the nineteenth century colonial imaginary. . . . The significance of the exhibition rests on the ability for a re-interpretation that

> goes beyond celebration or condemnation, it provides an opportunity for recoding the fragments of the historical unconscious. (Papastergiadis, 1994: 12)

Museums

If Gertsakis' installation conceptually 'recodes' the fragments of the historical unconscious, the final installation 'White Apron' – 'Black Hands' physically exemplifies such an approach. The work of King-Smith, Black and Huggins is not only museologically sound, in that it brings together oral histories, historic photographs, original documents and object research, the installation also consumes the visitor in a uniquely sensual, textural and emotionally encompassing experience. In the collective's re-imag(in)ing of the stories told by Aboriginal women domestic servants at the turn of the century, the intertwining of the personal archive and the institutional archive is so delicately and yet powerfully orchestrated that it makes one pause to reconsider whether or not 'objective' curation is an attainable, even desirable, option when material as sensitive as this is being presented.

Leah King-Smith would testify that it is not. Her richly layered photographs, which iconographically reunite Aboriginal women with their land and their artifacts, are the result of an intimate connection made between the artist and, in this case, the library. Inspired by her work with archival photographs held in the State Library of Victoria, King-Smith writes that her initial introduction to images of black domestic servants

> began as a standard research task, locating and compiling the Aboriginal photographs amongst the Library's enormous range of pictures. But uncovering those powerful images evoked, in time, such strong feelings in me that my work needed to evolve creatively in order to deal with the overwhelming feelings that were coming up. I was seeing the old photographs as both sacred family documents on the one hand, and testaments of the early brutal days of white settlement on the other. (Black, Huggins and King-Smith, 1994: 7)

King-Smith's double vision, her 'two-fold story of powerlessness and powerfulness', is perhaps the very story which needs to be told more openly in our museums today. It is in offering us visions, at once aesthetic and academic, that artists such as Gertsakis and Black Day Dawning have suggested a new paradigm for curatorial practice. Why not allow archival photographs to stand Janus-faced as symbols of both racial dominance and individual resistance? Why not encourage artifacts to reflect complex historical realities, at the same time allowing them to evoke personal responses to aspects of that history? Uniquely positioned as both artist and curator (and what, in fact, are the boundaries which separate one from the other?) these women, as self-proclaimed agents of interpretation, have quite purposefully melded personal histories with institutional histories. They have infused their objective readings with highly subjective reactions.

Treachery

Well, yes, we might say – these artists can do all this because they are artists. Theirs is a creative response full of interpretation and provocation. The works they produce are installations, after all, not exhibitions in the traditional museum sense.

And yet, it is these very distinctions – between exhibition and installation, curation and creation, presentation and dramatization – which are being called into question by artists such as Aleks Danko. Danko's 'Zen' installation used objects from the University of Melbourne's collection to create environments which, in Green's words, 'turn rigour into farce' (Green, 1994: 14). In blurring the lines between curation, installation and window dressing Danko asked his visitors to question how institutional structures work to construct hierarchies of aesthetic taste or historical importance. By working with irony and parody, as he has done in many of his works, Danko joins Gertsakis and Black Day Dawning as an artist-curator attempting to interrupt and then shift the codes of museum presentation.

While Gertsakis' 'treachery' comes with her investigations into the 'pictorial interstices' where images of the *other* had been allowed to slip – and Black Day Dawning's traitorous act was to revitalize the objects, images, and landscapes of the white-aproned, black-handed women whose stories are still alive today – Danko's artistic trespass is his rigorous irreverance toward accepted strategies of institutional interpretation. Danko knows and plays the game very well, but he always twists it. Green writes of this approach:

> The paradoxical combination of near-total museological professionalism with a fluid notion of identity is seen in the artist's extraordinary organisation and his erratic relationship to the more conventional signs of

authorship . . . The eroticism and hyper-alert jokes that saturate many of his works have the same deliberate effect of disorienting taste (Green, 1994: 4).

BEING THE ARTIST IN OUR OWN MUSEUMS

So, does The Artist and the Museum series go beyond telling 'hyper-alert' jokes and disorienting our taste; does it do more than direct our attention to the way we read symbols of identity and power; is it striving to expose other facets of Australian history that exist parallel, or in opposition to those embodied by collections held in trust by the country's major cultural institutions? I would say that it does indeed.

At one level, the very presence of these installations and the ideals to which they aspire – among other things: 'penetrating the mysteries of the classification systems of museum in a spirit of "institutional critique" ' (Gates, 1994: 30) – signals to the larger interpretive community that the potential of museum collections has barely been tapped. Whether in a more conventional way as historical resources and cultural archives, or along more radical lines as political tools, performative aids or installation pieces, the 'stuff' of museums is coming under newly inspired scrutiny.

On a deeper level, the more radical potential of these installations lies in their exposure of unconventional interpretive approaches. Manipulating diverse media and pursuing uniquely personal and political issues, the series artists are still linked by their common strategy of taking works out of a formal collecting body and re-institutionalizing them in the archives of memories and galleries of experience. These artists do not treat curation and criticism as objective, omniscient activities. Rather, their installations point to a mode of curatorial practice which is in the broadest sense transformational instead of being strictly interpretational. Their reading and re-presentation of pieces such as Nicholas Gysis' Love's Pilgrimage, a nineteenth-century carved emu egg, or an 1888 photo entitled: 'Believed to be a maid in the Cambell household', is derivative, inspired, quotational, contextual – oblique. The final installations draw our eyes to the spaces *between* the museum's lines rather than to the lines themselves.

My praise of The Artist and the Museum series is meant to be neither didactic nor prescriptive for those working in the more traditional spheres of curatorial activity. The dialogical and enunciative premises of the installations are certainly not transferable, nor appropriate, to many institutions which are in the process of readdressing their collections. Nevertheless, such major projects as this one curated by Gates at the Ian Potter Gallery make for essential reading in today's museologically complex world. For those involved in cultural institutions and in the discipline of museum studies, the conceptual intermeshing of 'crisis', 'museums' and 'treachery' cannot possibly go unnoticed. Nor can the exciting possibilities, which such interventions have exposed, go untried. Perhaps it is time we each became 'artists' in the museum and stepped back to reconsider our presence in the collections, but also our collections' presence *in us*.

BIBLIOGRAPHY

Aleks Danko: Zen Made in Australia (1994) (University of Melbourne Museum of Art).

Barnes, C. (1993) 'Relational space' in *Peter Cripps: Projects For Two Museums* (University of South Australia Art Museum, pp. 25–47.

Barrett-Lennard, J. (1993) 'Projects for two museums' in *Peter Cripps: Projects For Two Museums* (University of South Australia Art Museum), pp. 8–24.

Black, L., Huggins, J. and King-Smith, L. (1994) *White Apron – Black Hands* (University of Melbourne Museum of Art).

Elizabeth Gertsakis: Beyond Missolonghi (1994) (University of Melbourne Museum of Art).

Gates, M. (1994) 'The elevator' in *Aleks Danko: Zen Made In Australia* ed. (University of Melbourne Museum of Art: 28–34.

Green, C. (1994) 'Aleks Danko' in *Aleks Danko: Zen Made In Australia* (University of Melbourne Museum of Art: 2–15.

Papastergiadis, N. (1994) 'Greece at the crossroads', in *Elizabeth Gertsakis: Beyond Missolonghi* (University of Melbourne Museum of Art): 10–20.

Peter Cripps: Projects For Two Museums (1993) (University of South Australia Art Museum).

Writing about Art Exhibition Catalogues: A Literature Review

PNINA WENTZ
Centre for Information Management, Thames Valley University

Art exhibition catalogues vary greatly in scope, content and format: from lists providing basic information to extensive high quality publications containing primary source materials such as results of primary research, reproductions of works of art not illustrated elsewhere, interviews with artists, scholarly articles and critical essays. An exhibition catalogue, even a simple listing, may be of considerable value, as it may constitute the only record of the existence of the works of art on display and the only evidence that the exhibition took place. Thus, art exhibition catalogues can be considered as a primary research tool for the art historian, though to a varying degree. In addition, art exhibition catalogues provide a rich source for studying changing policies and practices of exhibiting works of art over a period of time, the changing theoretical framework and practices of writing about works of art, changes in the perceived role of the catalogue in relation to the exhibition it documents and changes in the methods of publishing and production which reflect contemporary cultural, economic and technological trends. This paper is a short review of published literature about art exhibition catalogues.

'Considering the huge expenditure incurred every year in publishing art exhibition catalogues it is remarkable how little attention has been given to the art exhibition catalogue per se', states Cannon-Brookes (1985: 22), and Burton comments that 'exhibition catalogues lack a critical theory' (Burton, 1977: 72). Searches of several on-line and printed sources support these statements. Most of the relevant literature identified offers short contributions, observations and insights borne out of experience and knowledge rather than extensive studies and research.

Cannon-Brookes outlines the history of catalogues of temporary art exhibitions by focusing on significant publications that constitute landmarks in the evolution of art exhibition catalogues from simple checklists, as they remained until the end of the nineteenth century, to massive lavishly illustrated volumes. Noting that 'The watershed in attitudes towards the function of exhibition catalogues was the publication in London of the commemorative catalogues of the Burlington Fine Arts Club from circa 1900', Cannon-Brookes (1985: 23), points out that these catalogues were not intended to be carried around the exhibition like checklists and were in fact published after the exhibition. He also draws attention to the increasing significance of the art exhibition catalogue as a medium for publication of research since the 1960s.

Docken Mount (1988) studied the increasing significance of art exhibition catalogues as a tool for publishing research in a specific area. She examined the evolution of exhibition catalogues of African art in the United States and their growth from slim checklists to large scholarly publications. Changes occurring over the years were traced through examination of form and content of selected catalogues and by studying review statements made by art historians about these catalogues. In addition, she analysed bibliographic citations in the journal *African Arts*, a major forum for scholarly writing on African arts in the United States. Her analysis reveals increasing use of exhibition catalogues as cited sources for scholarly articles in *African Arts*. Finally, she studied the sources and annotations included in a bibliographic guide to the literature of African art and was able to establish the prominence of exhibition catalogues among the cited sources.

Several other contributions offer some discussion of the different types and functions of exhibition catalogues, their structure and value as records of research and as primary sources of documentation and illustrations for art historical research (Burton, 1977; Davis, 1973; Leach, 1977; Neuheuser, 1988; Phillpot, 1973; Robertson, 1989; Smith, 1973). It is worth noting that most of these authors are art librarians who are concerned with the problems associated with the provision of exhibition catalogues as library documents. Exhibition catalogues are often described as 'fugitive' or 'elusive', most are not available through the book trade and may run out of print even before the exhibition has closed. Once obtained, further difficulties are presented by the great variety in format and structure of catalogues. Thus, the emphasis on the value and significance of exhibition catalogues in the writings of art librarians can be seen as an attempt to legitimize the efforts required to obtain catalogues and make them accessible to users through special indexes and bibliographic tools.

The most extensive study of this kind is the feasibility study on establishing a national collecting network for art exhibition catalogues, conducted by Smith and Jackson for ARLIS (Smith and Jackson, 1990). To overcome the problem of identifying and obtaining catalogues of temporary exhibitions ARLIS proposed setting up a network of art librarians based in art libraries throughout the country who will become local collectors. The main study concluded that a national collecting network based on local volunteers had no future viability. As part of the feasibility study a detailed questionnaire survey to identify the extent and the frequency of catalogue publishing in England between September 1988 and August 1989 was carried out. A total of 238 useful questionnaires were returned and by the end of 1989 the database contained 1,590 catalogue records, of which 76 per cent were fine arts catalogues.

Analysis of the database offers some interesting insights into the characteristics and publishing practices of art exhibition catalogues. Most catalogues in the database are on twentieth-century art and a very high proportion relates to the postwar period. Most catalogues are illustrated and a substantial proportion have some colour illustrations.

> The analysis of the database records would seem to confirm the unconventional and insubstantial nature of art exhibition catalogues. . . . 68% consisted of less than 32 pages, just over a quarter were cards/folded cards or single sheet lists. Other formats included sets of cards or postcards in folders, posters, wallcharts. (Smith and Jackson, 1990: 41)

The survey reveals the non-commercial nature of most catalogue publishing. Most catalogue producers appeared to publish relatively few titles, about half produced only one title during the survey period and less than 10 per cent produced more than six catalogues, with the National Galleries identified as the most prolific producers (the South Bank Centre 28 catalogues and the Tate 25). Co-publishing with commercial publishers was rare; only 4 per cent of the catalogues on the database were co-published, but these were mainly joint gallery ventures with only a small number involving a commercial publisher. Smith and Jackson refer to Thames and Hudson, Weidenfeld and Nicolson, and Lund Humphries as partners in co-publishing.

In the United States co-publishing is much more widespread, as Smith and Jackson note 'the "book to accompany an exhibition" is becoming increasingly popular . . . co-publishing of catalogues has become big business with firms like Abrams, Rizzoli and the major university presses . . . competing for licensing of books and catalogues' (Smith and Jackson, 1990: 6). A practical 'how-to' guide to negotiating co-publishing arrangements is offered by Brown, director of publications at The Museum of Modern Art, New York, who provides a step-by-step checklist of points to consider at each stage of the process (Brown, 1993).

In spite of the advantages of co-publishing in providing professional production, marketing and distribution facilities, significant concerns are expressed by several contributors. Cannon-Brookes points out that the association with commercial publishers has resulted in 'increasing independence of the exhibition catalogue from the exhibition which it ostensibly served' (Cannon-Brookes 1985: 24). Catalogues became increasingly more ambitious in scope and size as 'the design and production of the art exhibition catalogue became ever more influenced by the demands of book production and marketing'. Many catalogues have become too heavy to carry around the exhibition and contain 'too much information to be digested during a single visit'. Cole argues that such catalogues do not benefit the majority of visitors who need 'a concise well-written guide to an exhibition that the visitor could read while in the galleries' (Cole, 1993: 49). He notes that

> recently some of these blockbuster catalogues have been leading a double life. They are copublished by commercial art publishers who disguise them as coffee-table books and market them widely. That an exhibition catalogue can be repackaged with such ease demonstrates how the traditional line between book and catalogue has now become blurred. (ibid.)

Acknowledging the economic realities which may encourage co-publishing, Cannon-Brookes points out that the demands of book publishing may result in a subtle change in priorities and 'an element of doubt begins to arise as to whether the catalogue is intended to accompany/commemorate the exhibition, or whether the exhibition has been arranged to market the catalogue' (Cannon-Brookes, 1985: 26).

Other contributions to the discussion on museum publishing include the papers prepared for a seminar run by the National Museums of Scotland in 1987 (Calder, 1988). The topics covered include planning, copy-editing, design, production and marketing of museum publications. A more recent contribution outlines the potential of new technology to create new types of catalogues. Baker (1993) reviews the development of the Micro Gallery project at the National Gallery, pointing out that the same technology (a custom adapted HyperCard stack) has been used for the production of a new printed catalogue. In addition, a CD-ROM version of the system has been issued to make the catalogue widely available in electronic form.

Although there is significant literature on the general subject of the relationship between writing about art, the art market and the ethical issues involved[1], not much has been published specifically about writing in art exhibition catalogues. Duret-Robert (1989) discusses the ethical issues in compiling contemporary *catalogues raisonnés* and the implications for the art market, and Umberto Eco, in a satirical essay (Eco, 1984), surveys the pressures and pitfalls of writing for art exhibition catalogues.

Remarkably little is to be found in the published literature on the role and content of exhibition catalogues of contemporary art. An exception is an article by Morgan (1991), who contrasts two different approaches. His first example is the catalogue of the Whitney Museum of American Art that documents the museum's 1987 biennial exhibition. The catalogue contains many high quality colour photographs but the only texts are two brief descriptive overviews and a foreword by the museum's director. With minimal text, the contextual reading of the exhibition which is manifested by the way the works are displayed is lost in the isolation of the photographs on the pages of the catalogue. Morgan argues that text is necessary as 'language is the glue that holds the impact of these signs together' (Morgan, 1991, p. 343) in the catalogue. Although a theoretical position is implicit in the curatorial selection, it would be more credible to articulate a theoretical position within the catalogue rather than 'husbanding critics outside the perceived establishment of the museum to legitimate the concerns of the curatorial staff' (ibid.).

A contrasting example is the catalogue of the exhibition *Damaged Goods* at the New Museum of Contemporary Art in New York in 1986. The title highlights the theoretical premise of the exhibition which is further discussed in critical theoretical essays and in statements and documentation provided by the participating artists. The texts are accompanied by a detailed bibliography 'to further legitimate the claims that support the curatorial intentions'. Black-and-white photographs are used and are spread throughout the various essays. Morgan believes that 'the catalog maintains a sense of its own self-containment to the point of creating a breach between the text and the exhibition itself' and that 'in one sense the exhibition was superfluous' (ibid.). He argues that although theory and discourse are needed, the 'over-determination of the exhibition by way of the catalog cancels any possibility of an internalized discourse emanating through one's direct contact with and reflection on the works themselves' (Morgan, 1991: 344)). Exhibition catalogues should not distance the visitors from the experience.

Turning to the primary materials, i.e. to the exhibition catalogues, we discover a rich, almost untapped source for study. From these we can identify curatorial policies and practices of exhibiting works of art over a period of time, the theoretical framework and practices of writing about works of art, discussions on the role of museums and galleries and the perceived role of the catalogue in relation to the exhibition it documents. Such a study, however, is beyond the scope of this literature review, but one on which I hope to report in the future.

NOTES

1 See for example Bowness (1989) for a discussion on the role of writing about art in 'the modern artist's rise to fame' (p. 25), and a more extensive discussion in Carrier (1987) and Duncan (1993).

BIBLIOGRAPHY

Baker, C. (1993) 'A marriage of high-tech and fine art: the National Gallery's Micro Gallery project', *Program*, 27(4): 341–52.

Bowness, A. (1989) *The Conditions of Success: How the Modern Artist Rises to Fame* (London: Thames and Hudson).

Brown, O. (1993) 'Partners in print: a copublishing how-to', *Museum News*, 72(4): 46–7 and 61–2.

Burton, A. (1977) 'Exhibition catalogues' in P. Pacey (ed.) *Art Library Manual: A Guide to Resources and Practice* (London: Bowker), pp. 71–86.

Calder, J. (ed.). (1988) *Museum Publishing: Problems and Potential* (Edinburgh: National Museums of Scotland).

Cannon-Brookes, P. (1985) 'The evolution of the art exhibition catalogue and its future', *Art and Artists*, 220: 22–7.

Carrier, D. (1987) *Artwriting* (Amherst: University of Massachusetts Press).

Cole, B. (1993) 'Catalogues of abuse', *Museum News*, 72(4): 48–9.

Davis, A. (1973) 'Exhibition catalogues: importance, format, bibliography', *ARLIS Newsletter*, 16: 13–18.

Docken Mount, S. (1988) 'Evolution in exhibition catalogues of African art', *Art Libraries Journal*, 13(3): 14–19.

Duncan, C. (1993) 'Who rules the art world' in C. Duncan, *The Aesthetics of Power: Essays in Critical Art History* (Cambridge: CUP), pp. 169–88.

Duret-Robert, F. (1989) 'La dictature des cataloguistes', *Connaissance des Arts*, 448: 131–43.

Eco, U. (1984) 'Wie man das Vorwart für einen Ausstellungskatalog schreibt' [How to write an introduction to an exhibition catalogue], *Du*, 5: 6–12.

Leach, E. (1977) 'Sales catalogues and the art market' in P. Pacey (ed.) *Art Library Manual: A Guide to Resources and Practice* (London: Bowker).

Morgan, R. (1991) 'The exhibition catalog as a distancing apparatus: current tendencies in the promotion of exhibition documents', *Leonardo*, 24(3): 341–4.

Neuheuser, H.P. (1988) 'Ausstellungskataloge als spezifische Publikationsform' [Exhibition catalogues as a specific form of publication], *Bibliothek: Forschung und Praxis*, 12(3): 241–62.

Phillpot, C. (1973) 'Exhibition catalogues: notes on categories, use and standardization', *ARLIS Newsletter*, 16: 10–12.

Robertson, J. (1989) 'The exhibition catalog as source of artists' primary documents', *Art Libraries Journal*, 14(2): 32–6.

Smith, G. (1973) 'Exhibition catalogues in the United Kingdom: values and problems', *ARLIS Newsletter*, 16: 6–9.

Smith, G. and Jackson, L. (1990) *ARLIS/UK Eire National Collecting Network for Art Exhibition Catalogues*, British Library research paper 87 (London: British Library).

The Museum of Women's Art: Towards a Culture of Difference

BEVERLEY BUTLER
Department of Conservation and Museum Studies, University of London, Institute of Archaeology

Ours is a culture which refuses to celebrate the achievements of women. Having been present at the official launch of the new Museum of Women's Art (MWA) I was disappointed, though not surprised, to find that when the news reached the pages of our 'liberal' press the emphasis had shifted from that of celebration to comment and analysis based upon frames of reference which exclude women.

Those dualisms which underpin Western culture provided convenient and self-fulfilling critiques which 'proved' the weaknesses inherent in separatist activity and the subsequent retreat into ghettoization (Anon, 1994: 19). Constructs such as 'quality', like that of 'genius', were reaffirmed as sole criteria for judging artistic creativity; criteria, it is suggested, the MWA may fail to uphold and could instead be subsumed by a form of 'positive discrimination' based upon 'mere gender correctness' (Morrison, 1994: 18).

How could women sing their own praises? Safely defined as the 'other', all creativity and experience is discredited. The debate has proved nihlist but effective enough to mitigate any sense of achievement. All of which ensures that 'our' culture remains unrepresentative.

Only by seeking new frames of reference can we explore the tension between critique and vision and recapture the spirit of celebration. With this in mind I intend to draw upon the work of the French psychoanalytical philosopher Luce Irigaray and to place the MWA in the context of Irigaray's own vision – a culture of difference.

THE PROJECT: AN INVITATION TO CREATE A MUSEUM OF WOMEN'S ART

> It is quite simply a matter of social justice to balance out the power of one sex or gender over the other by giving it, or giving back, subjective and objective fights to women. (Irigaray, 1993a: 80)

Monica Petzel, MWA's honorary secretary said, 'to build their own history and create their own museum. It would be a powerful thing for women to do' (Weale, 1994: 6).

The Museum of Women's Art has made a symbolic intervention in our contemporary cultural arena. On a conceptual level it promises space to study, exhibit and nurture the artistic creativity of womankind. Still in its embryonic form the museum has yet to establish a permanent site. Its genesis has been truly utopian [no place].

The process of becoming has involved the formation of an intellectual rationale and the creation of an identity in the public domain. In this the MWA has harnessed and consolidated those histories and visions which make claim to some kind of women's geneaology.

Feminist critiques in art history (Greer, 1979; Nochlin, 1973; Parker and Pollock, 1981) have informed the museum mission and intellectual agendas. Similarly, the campaigns of the Guerrilla Girl's (Gablik, 1994: 6–11) which exposed the shameful bias of existing galleries and museums in the United States have provided a precedent for the MWA's own survey of national museums and galleries in Britain (see figure). In Washington's Museum of Women's Art the MWA found a mother/sister figure to emulate.

Unique in Europe, the museum is a creative exploration of the possibilities of a visual incarnation of women's culture. An economic base is being developed through the acquisition of charitable status and ongoing appeals for funding, while an influential network is manifest in the MWA supporters/advisers, specially selected from a variety of 'cultural' spheres including the worlds of art, academia, museums, galleries and feminism.

Gallery/museum	*Total collection*	*Works by women in collection*	*Works by women on display*
Tate, London	15,000	1,200	50
Castle Museum, Nottingham	5,300	143	7
Walker, Liverpool	3,292	92	13
National Portrait Gallery, Edinburgh	2,820	56	8
National Gallery, London	2,214	11	3
Birmingham City Gallery	1,020	61	3
	29,646	1,563	84

Source: information provided by galleries

* Museums and galleries display only a small proportion of the work by women in their collections.
* It has been established that there is a large pool of work by women which could be made available to MWA.
* Works by women represent a very small proportion of the total works held in most British galleries.

Figure 11.1: Works of art in six British galleries, from publicity materials from MWA.

IDENTITY SPACE: SACRED SITES AND SINGING BIRDS

> Mouseia were sacred sites where birds sang – thought to be manifestations of the gods
> (Larrington, 1992: 86)

> To construct and inhabit our airy space is essential. It is space of bodily autonomy, of free breath, free speech and song, of performance on the stage of life.
> (Irigaray, 1993b: 66)

The mythologies of the ancient world are fundamental to the identity of our contemporary (Western) culture. It is in whose image we define ourselves, construct our genealogies – our limitations and visions – and create our social order. Our language and thought reflect these beginnings.

Female genealogies are absent: a severing of the link with the goddesses of prehistory's matriarchal culture has left women bereft of their, our, cultural lineage. The museum has played a symbolic role. It was on this stage, this site, that Gaia was usurped of sacred space and sacred status. The temple at Delphi, originally dedicated to Gaia, was appropriated by Apollo who symbolically silenced womankind when he bid the muses to sing in his voice (Larrington, 1992: 85–6).

The need for women to create a symbolic home of their own has become a recurring theme in attempts to give voice to female creativity. Virginia Woolf spoke up for women writers in a polemic which can be extended to all the creative endeavours of women (Woolf, 1929). Irigaray's own concept of 'identity space' (Irigaray, 1993a: 83) takes this notion of sexuate space forward to the vision of difference. She sees identity space as crucial to 'converting' the male/female dichotomy, 'she [Irigaray] warns against displacing the male/female binary before the female side has acceded to identity and subjectivity' (Whitford, 1994a: 13).

Within the framework of patriarchy both the conceptualization and attempts to create such space have degenerated to bipolar discourse – with women fearing the 'marginalized' position. Such debates surrounded the creation of the Women's Pavillion at the Centennial Exhibition in Philadelphia in 1876 and are true of attitudes (including those of some feminists, see Morrison, 1994: 18) to the MWA. Lack of consensus is interpreted as 'women's inability to organise themselves and agree on what they want' (Irigaray, 1994: 12). The critiques depress where the vision excites. Irigaray poses a different question and asks 'how could they [women] unite when they have no representation, no example of such an alliance' (Irigaray, 1994: 12).

The MWA has the potential to capture Irigaray's vision, to become an identity space and an example of 'united women', while 'the museum' in its symbolic institutional form could become a significant part of female genealogies. The museum, like the women, could then begin to (re)search its own histories and to rediscover its sacred status and its tones of free speech and song.

RECLAIMING THE MADONNA: DIVINE WOMEN

> In the museum there is a statue of a woman who resembles Mary, Jesus's mother, sitting with the child before her on her knee, facing the observer. I was admiring this beautiful wooden sculpture when I noticed that this Jesus was a girl! That had a very significant effect on me, one of jubilation – mental and physical . . . Standing before this statute representing Mary and her mother, Anne, I felt once again at ease and joyous, in touch with my body, my emotions and my history as a women. (Irigaray, 1993a: 25)
>
> Reclaiming the Madonna, the first public manifestation by the MWA, is a mixed show which attempts to redress the balance of those placid icons by showing motherhood from the inside; as an activity rather than as a consoling idea. (Kent, 1994: 49)

For Irigaray, measuring oneself against the divine is the starting point for women to define their own identity (Whitford, 1994a: 13). It is apt that the exhibition selected to launch the MWA – 'Reclaiming the Madonna' – took the concept of the divine woman as 'subject'. The exhibition which originated at the Usher Art Gallery, Lincoln, was installed in the Economist Tower, London; the first of three events planned during the museum's developmental stage.

The vision mirrors that of Irigaray's concept of the mother/daughter dynamic, the relocation of women's culture with the divine (Irigaray, 1994: 8–14). Women are objectified within our culture (within our museums) but there are too few art and artefacts to symbolize their experience. Where images do exist in the public arena (whether photographs, sculpture, advertisements) they are usually produced by men for the male gaze rather than for women to gain some sense of their own history.

It is rare for women to visit any public venue, any space public which can communicate symbolic attachment to the female genealogy unmediated by male desires. The question is posed by some critiques, 'is there such a thing as women's art, women's culture? (Morrison, 1994: 19). We should remember that museums can be a fertile space in which culture can be *created*, not just collected, preserved, conserved, exhibited and interpreted.

NEW VISION: THE SYMBOLISM AND CELEBRATION OF DIFFERENCE

> Claims that men, races, sexes, are equal in point of fact signal a distain or a denial for a real phenomena and give rise to an imperialism that is even

> more pernicious than those that retain traces of difference . . . Sexual difference represents one of the great hopes for the future . . . [it is to be found] . . . in the access the two sexes have to culture.
> (Irigaray, 1993b: vi)

> To seek to cage up within the domestic setting something that has always flowed uncontained is like turning free-soaring, rapture, flight into parchments, skeletons, deathmasks. (Irigaray, 1993b: 42)

The agenda the MWA is that of cultural restitution. A symbolic return of art and artefacts, of identity, of genealogy in the name of sexual difference. Ours is a culture which *will not* celebrate the achievements of women. Ours is a culture which *cannot* celebrate the achievements of women. Our culture remains unrepresentative, exclusive and unresponsive to the Other. The MWA symbolizes one attempt to redress the balance to make 'our' culture, ours.

The museum has already made a significant intervention, one which has redefined existing cultural institutions in many people's minds. A relationship of cultural exchange which transcends traditional notions of equality is demanded. It is a chance for museums to reaffirm their own commitment to social justice. It is a positive moment in museum genealogy, an opportunity for them to participate in the celebrations, rituals and festivals of difference and to extend a promise, a voice and a cultural inheritence to all groups who have suffered similar marginalization.

Museums are an arena of contestation on whose stage, or sacred earth, our identity is (re)created and all our symbolic hopes for cultural restitution are made. It marks a new relationship between peoples, sexes and cultural exchange. For Irigaray the *culture of difference* denounces the idea of the 'project' and instead releases all elements into a new creation and contestation,

> if anything divine is still to come our way it will be won by abandoning all control, all language, and all sense already produced, it is through risk, only risk, leading no one knows where, announcing who knows what future, secretly commemorating who knows what past. No project here.
> (Irigaray, 1993b: 53)

Only via these new visions can the spirit of celebration and achievement of the 'Other' be translated and articulated within our culture. The MWA can be interpreted as representing this type of symbolic intervention and of giving voice to women's creativity with all the hope and discovery this can yield, 'silently their song irrigates the world of today of tomorrow, of yesterday . . . goes on to nothing, or something greater than anything we now have' (Irigaray, 1993b: 53).

BIBLIOGRAPHY

Anon (1994) Guardian Leader, 'A prize that is second best', *The Guardian*, 15 August: 19.

Gablik, S. (1994) 'You don't have to have a penis to be a genius', *Women's Art Magazine*, 60: 6–11.

Greer, G. (1979) *The Obstacle Race*, (London: Weidenfield & Nicholson).

Irigaray, L. (1993a) *Je, Tu, Nous: Towards a Culture of Difference*, trans. Alison Martin (London: Routledge).

Irigaray, L. (1993b) *Sexes and Genealogies*, trans. Gillian C. Gill (New York: Columbia University Press).

Irigaray, L. (1994) *Thinking the Difference: For a Peaceful Revolution*, trans. Karin Montin (London: Athlone Press).

Kent, S. (1994) 'Mother Courage', *Time Out*, 29 June: 49.

Larrington, C. (ed.) (1992) *The Feminist Companion To Mythology*, (London: Pandora).

Morrison, B. (1994) 'Buried Treasure?' *Independent on Sunday*, 3 July: 18–20.

Nochlin, L. (1973) 'Why have there been no great women artists?' in E. Baker and T. Hess (eds) *Art and Sexual Politics* (London: Collier Macmillan).

Parker, R. and Pollock, G. (1984) *Old Mistresses: Women, Art and Ideology*, (London: Routledge and Kegan Paul/Pandora).

Saunders, K. (1994) 'Movement in the wrong direction,' *Sunday Times*, 26 June: 10.

Stringer, R. 'Gallery will put women in a class of their own', *Evening Standard*, 21 July: 8.

Weale, S. (1994) 'A brush with fame', *Guardian Women*, 14 June: 6–7.

Whitford, M. (ed.) (1994a) *The Irigaray Reader* (Oxford: Blackwell).

Whitford, M. (1994b) 'Woman with attitude' *Women's Art Magazine*, 60: 15–17.

Woolf, V. (1929) *A Room of One's Own* (London: Hogarth Press).

Health Matters in Science Museums: A Review

RAJ KAUSHIK
Department of Museum Studies, University of Leicester

Health matters? Who could deny it in the 1990s particularly when the diseases of modern society are attributed to life style and behaviour. The clients of medicine are no longer simply people who are ill, but potentially all of us. The responsibility for maintaining health and preventing illness now appears to rest with the individual. In these circumstances, there is an obvious and increasing need to promote health literacy at national level.

Since the beginning of the present century, science museums have been playing a socially responsible role with regards to health issues in society. For example, in 1908 the American Museum of Natural History opened an exhibition on tuberculosis to deal with the overwhelming response of a concerned and frightened public. Today, the AIDS epidemic is impacting on societies in a similar way and prompts museums to react accordingly. It is heartening to see that science museums are again identifying the right priorities. Recently, two exhibitions specially aimed at adult visitors opened in London: 'Science for Life' at the Wellcome Centre for Medical Science in January 1993 and 'Health Matters' at the Science Museum in June 1994. In the 1990s, developing a gallery for mainly adults sounds idiosyncratic. In fact, in the 'Science for Life' exhibit, two information boards (mounted on walls) entitled 'Important Notice to Visitors' read as:

> 'Science for Life' has been designed with an adult audience in mind. Visitors bringing small children to the exhibition are requested to keep them under close supervision at all times, for reasons of safety and to minimise the risk of damage to sophisticated exhibits.

Despite some similarities, for example theme, financial aspects and target audiences, the two exhibits adopt a radically different approach. On one hand, Health Matters aims to highlight the changes in medical theory and practice which make our perception of medicine radically different from that of our grandparents. On the other hand, Science for Life intends to enthuse and educate visitors about the workings of the human body in health and disease, and to help them share the excitement that scientists have in their creative endeavour in discovering the mysteries of life. While Health Matters relies heavily on medical technology and its gadgets, Science for Life throws light on the nature and processes of science.

Like many recent exhibits, Health Matters is a result of partnership between sponsors (SmithKline Beecham, Action Research, The British Diabetic Association, British Heart Foundation, Medical Research Council, Multiple Sclerosis Society and the Wellcome Trust), designers (Jasper Jacob), audio-visual experts (Triangle Two), interactive designer (Dunne and Raby) and curators (Ghislaine Lawrence and Tim Boon). Only visitors were not made partners in the initial stages. An estimation of the potential visitors for the exhibit was predicted in a study (Bicknell, 1995) which gives the impression of having been conducted in order to prompt industrial houses to collaborate.

While front-end evaluation was conducted for the Science for Life exhibit, Health Matters is purely the curators' brainchild. In response to my question on the criteria for selection of the details of subject matter, Dr Tim Boon said:

> It may sound very arrogant to say that our view was, I think, very much informed by our backgrounds as historians of medicine . . . social historians in general as well as pretty experienced museum curators about what was needed in the gallery.

In essence, the attitude of the curators can be explained as: 'Look visitor! historically, these changes in medical theory and practice and these changes in our perception about medicine are important in the twentieth century. You must know about these.' In contrast, visitors' views can be seen in the visitors' comment book placed near the exit of the gallery. Here, while most visitors write their impression about the museum as a whole, about Health Matters there are mixed reactions. However, critical observations are very enlightening. A few visitors' reactions will be used in this paper at appropriate places (Visitors' Comment Book, January 1995).

The exhibit Health Matters is comprised of three sections: 'The Rise of Medicine', 'The Rise of Health', and 'Science in Medicine'. The Rise of Medicine focuses on the technology and drugs which have become essential parts of modern medicine. Aside from caring for individual patients, there is another perspective to look at with health – the examination of rates of health and illness in populations. Section two, The Rise of Health, follows this 'epidemiological' approach and employs four interactives to explore some of the health indicators. In the early decades of the present century science and medicine were definitely two different worlds. But, today it is not so. Section three, Science in Medicine, highlights the wedlock of science and medicine and tells about our understanding of diseases which have been detected recently, such as AIDS, cancer and coronary heart disease.

The content of labels and their presentation is really excellent. The technical terms and the noteworthy events in the texts are highlighted in contrasting colours and explained in freehand in margins. The finding of perspective research suggests that what is learned and remembered from a text can be influenced by giving readers different perspectives about what they are going to read (Anderson, 1983). The style followed here really gives visitors a certain perspective that, in turn, influences cognitive processing by (1) grabbing attention, (2) acting as cues for prior knowledge relevant to a given topic, (3) accentuating the relationships among the concepts and facts, (4) suggesting analogies, and (5) providing retrieval cues for subsequent recalls. The potential effects of the technique employed here encourages visitors to read labels. As a result, comparatively more visitors can be seen reading labels intently than in any other galleries employing traditional textbook-type labels. Additional information about the various sections of the exhibit is provided on laminated sheets.

Instead of using a variety of interpretative techniques, mainly projection videos (video that projects image on a big screen) have been used in the exhibit. For example, in the first section, eight projection videos have been employed. The use of projection videos on a large scale creates several problems. First and perhaps the most severe problem is the low level of illumination in the gallery. Second, continuously running films all around on big screens depend on mainly two human senses, that is eye and ear, and thus, obviously, discourage conversation and social discourse among visitors. This is not desirable in the light of recent findings that learning in a museum does not primarily occur as a result of the interaction between individual and the exhibit but is also facilitated by social interactions among visitors at exhibits (Diamond, 1986; Blud, 1990). Third, the monotonous experience everywhere may also be boring, as is obvious from the following comment of a young visitor:

> I wish that they had some models that worked when you pushed a button and less tapes. They boring! Yes, playing with models is fun.

Participatory and interactive exhibits are widely recognized as viable antecedent for significant learning to occur. There are only a few participatory exhibits in the gallery. When I visited, many adults and children tried the 'health thermometer' and 'measuring health' interactive exhibits but did not always achieve the point of the activity. In order to create a successful interactive experience for visitors, the operation should, preferably, be so simple that even an unaided child could comprehend it. This cannot be said for the interactive exhibits presented here. The darkness in this section makes the situation even worse, as appears from the remark of a graduate student biology):

> Again very informative, especially, the medical history section. The epidemiology 'cone' should be better lit. In addition, the survey it housed was inaccurate about height, weight etc.

Most labels are placed at varying heights from one and a half feet to six feet above the ground level. For a physically fit person it is not comfortable to reach the lowest height to read the label, whereas the wheelchair-bound person certainly can not reach the highest. At many places in the gallery, there is less than optimum illumination level which creates further barriers for partially visually impaired people. It is surprising, but true, that in Health Matters the needs of the disabled are not taken care of.

Furthermore, in relation to content the issues of disability have been ignored bluntly. Basically, health and disability are the two sides of a coin. For example, on one hand a rare form of tuberculosis, if not diagnosed timely, can leave a person deaf. On the other hand one can lead a normal life if he or she gets treatment on time. There are approximately 6.5 million disabled people in Britain. Unfortunately, most of us perceive them as signifiers of ugliness, depravity or malignancy. Indeed, lots of people fight tooth and nail to stay out of a wheelchair. For them, to accept one is literally equivalent to giving up. In a technology-oriented gallery on health matters, one can genuinely expect to know more about the impact of medical technology on the lives of disabled people. By accentuating the essential link between health, disability and technology the exhibit would not have only catered to the needs of a huge population but also could have initiated pioneering work towards the end of appropriating culture, as our (discriminating) culture shapes what is understood by society.

One of the main strengths of the exhibit is that it is knitted around people. On one hand it talks about Alexander Fleming (discoverer of penicillin) and on the other hand about Moreen Lewis, the first home user of a kidney machine. Often, it gives a feeling of the prevailing social and circumstantial context of the period of the presented objects. For example, it highlights that iron lungs were made in a car company because of reduced demand for cars at that time. At many places, simple and familiar objects, such as a new shirt in a polythene pack, a rosebush. mashed potato, a car wheel and so forth, are placed, which really invoke curiosity among visitors to know more about why these things are there.

In family health matters, women play an important role. 'Although health is a man's single most important asset, he still prefers to abdicate responsibility for it to women,' says John Illman, a noted health reporter in the *Guardian* (18 January 1995: 15). Health Secretary Virginia Bottomley also makes the same point when she remarks:

> Who is it who decides which doctor to use? Who decides if the family has a healthy diet? Who decides whether the children should be immunised? Who encourages the family to take exercise and gives an example on smoking and alcohol? In disease prevention and health promotion, it is woman who are the key to success. (Quoted in the *Guardian* section 2, 18 January 1995: 15)

The crowds in the waiting-rooms of surgeries validate the informed views of Illman and Bottomley. Women are said to visit the doctor 50 per cent more than men do. So logically, a venture on health matters ought to consider women as its primary target audience and make profuse use of research on women's ways of learning and knowing (Kimura, 1992; Hill *et al.*, 1990; Belenky *et al.*, 1986) in its content, display and interpretation techniques. But, in Health Matters, the contrary can be observed.

The exhibit portrays science with a masculine character. Mostly, the role of male scientists have been projected prominently. Role models play a very important part in forming women's attitude to science (Hill *et al.*, 1990). In the exhibit, women are either patients or at the most nurses. The contribution of women in the development of medical science has not been discussed even where there are so many opportunities. For example, in X-ray discussion the work of Madam Curie could and should have been projected. It was Madam Curie alone who invested her efforts during wartime to bring the benefits of X-rays to people (Pflaum, 1989). As a further example, in DNA synthesis the work of Rosalind Franklin, whose research on X-ray diffraction of DNA crystals helped Francis Crick and James Watson to deduce the chemical structure of DNA, has been ignored.

In contrast, in the Science for Life exhibit, although Christina Ballinger (1993), deputy editor of *Museums Journal*, observes the 'masculine' image of science, the situation is not as bleak as in Health Matters. In fact, one can see female research graduates actually working in the demonstration area. Bright colours and spectacular audio-visual effects make the exhibition an exciting learning environment. About labels, Ballinger (1993) rightly points out that the hard and militant language is employed in the labels, but, in my view, supporting graphics in the labels make the task of visitors easy at the interactive exhibits. Labels, if written in the style used in Health Matters, would certainly have been more useful for visitors.

The exhibit Science for Life is mainly comprised of eight sections: 'Exploring the Body'; 'A Sense of Scale'; 'The Cell'; 'The Nature of Scientific Research'; 'Demonstration Area'; 'Major Advances'; 'Unanswered Questions'; and 'Funding'. Each section includes topics of general as well as academic interest. An effort is made to present the subject matter as plainly as possible. For example, at the entry of 'The Cell' exhibit a sense of scale is given by presenting a needle in varying magnification from four to one million times. Walking through a cell magnified a million times is an indelible experience. Most importantly, the exhibit gives a feeling about the provisional nature of science, about the incompleteness of science, and about the limitations of science and scientists.

It is really a difficult task to decide which of the two exhibits is better. While academic and scholarly people find Health Matters very informative and impressive, teenagers, adolescents and general public do not feel so. From the point of view of the general public, a visitor, Margaret from Holland, makes a decisive conclusion in the following words:

> (For Health Matters) Too messy, confusing, mixed up, too much put in a small place, no clear line, while there is obviously a lot of money spent on it. I can recommend 'Science for Life' in the Medical Science Building, Euston, much more.

The exhibit Science for Life has been conceptualized by Dr Peter Williams, director of Wellcome Trust 1969–91, and designed and developed by Met Studio Limited, London. Dr Laurence Samaje, who directed the project, Dr Bridget Ogilvie and Dr Stephen Emberton infused their efforts to make the exhibit a widely appreciated-reality. The exhibit has already won several prestigious awards, including the Museum of the Year Awards and the Lighting Design Awards. Also, the Fifth Design Week Award for the Exhibition Design 1994 has been awarded to Met Studio Limited for the Science for Life exhibit.

LESSON TO LEARN

Today, mounting on an exhibition without taking into account the visitors' point of view (epistemological commitments) is taking a gamble. From the above comments, it seems that visitors are very clear in their minds. Given a few alternatives, they can tell what is and what is not acceptable to them. In order to create a successful exhibition there is an increasing need to explore a compatibility between curators' and visitors' epistemological commitments. This can only be done by taking into account existing research findings and also the needs and interests of potential visitors.

BIBLIOGRAPHY

Anderson, R.C., Pichert, J.W. and Shirey, L.L. (1983) 'Effects of the reader's schema at different points in time', *Journal of Educational Psychology*, 75(2): 271–9.

Ballinger, C. (1993) 'One step beyond', *Museums Journal*, 93(10): 32.

Belenky, M.F., Clinchy, B.M., Goldberger, N.R. and Tarule, J.M. (1986) *Women's Ways of Knowing* (New York: Basic Books).

Bicknell, S. (1995) 'Here to help: evaluation and effectiveness' in Eilean

Hooper-Greenhill (ed.) *Museum, Media, Message* (London: Routledge), 286–8.
Blud, L.M. (1990) 'Social interaction and learning among family groups visiting a museum', *International Journal of Museum Management and Curatorship*, 9: 43–51.
Diamond, J. (1986) 'The behaviour of family groups in science museums', *Curator*, 29(2): 139–54.
Hill, O.W., Pettus, W.C. and Hedin, B.A. (1990) 'Three studies affecting the attitude of blacks and females toward the pursuit of science and science related careers', *Journal of Research in Science and Teaching*, 27: 289–314.
Kimura, Doreen (1992) 'Sex differences in the brain', *Scientific American*, 267(3): 81–7.
Pflaum, R. (1989) *Grand Obsession: Madame Curie and Her World* (New York, London and Sydney: Doubleday) pp. 183–229.

CALL FOR PAPERS FOR FORTHCOMING VOLUMES

The topic chosen for *New Research in Museum Studies* volume 7 is *Museums and Popular Culture*. The volume will concentrate upon the chosen topic, but may also include papers which deal with other aspects of museum studies. The editor would be pleased to hear from any worker in the field who may have a contribution. Any ideas for papers, or completed papers for consideration, should be sent to Professor Susan Pearce, Department of Museum Studies, 105 Princess Road East, Leicester LE1 7LG.

Contributions to the reviews section, including reviews of conferences and seminars, videos, exhibitions and books, or any other museum event, are similarly welcomed, and should be sent to Dr Eilean Hooper-Greenhill at the above address.

EDITORIAL POLICY

New Research in Museum Studies is a referred publication and all papers offered to it for inclusion are submitted to members of the editorial committee and/or to other people for comment. Papers sent for consideration should be between 3,000 and 10,000 words in length, and may be illustrated by half-tones or line drawings, or both. Prospective contributors are advised to acquire a copy of the *Notes for Contributors*, available from the editor, at an early stage.

New Research in Museum Studies has a policy of encouraging the use of non-sexist language, but the final decisions about the use of pronouns and so on will be left to individual authors. However, the abbreviation ‘s/he’ is not to be used.

The editor wishes it to be understood that she is not responsible for any statements or opinions expressed in this volume.

Notes on Contributors

Philip Doughty is Head of the Sciences Division in the Ulster Museum, Belfast. A geologist by background, his museum career began in local authority museums in Scunthorpe. He moved to the Ulster Museum in its major expansion phase and became its first Keeper of Geology in 1970. His main interests have been in the professional development of curation, particularly in collection welfare, standards development, documentation and the re-establishment of science in the cultural life of the UK.

Simon J. Knell joined the Department of Museum Studies, University of Leicester, in 1992, where he specializes in science curation and preventative conservation. He graduated from Leeds University in 1978 with a science degree in Geography, completed a Master's in Pollution and Environmental Control at Manchester University in 1982 before becoming a student on the course he now teaches. Knell worked in museums in Leeds, Manchester and Leicester before becoming Travelling Geology Curator for the Area Museum Service for South-Eastern England. Before taking up his current post he was Keeper of Natural Science at Scunthorpe Museums Service. He has been an active member of the Geological Curators Group and was for a number of years its secretary. He has also been Chairman of the Yorkshire and Humberside Collection Research Unit, a member of the Geological Society Conservation Committee and is on the Editorial Board of *Geology Today*. He has a passionate interest in the status and functioning of geology in museums; he is currently completing a PhD and book on the collecting of palaeontological material in the early nineteenth century. Knell has produced *Geology and the Local Museum* with Michael A. Taylor in 1989 and has published numerous articles on museums and the local geologist in *Geology Today* and elsewhere.

He has recently edited a *Bibliography of Museum Studies*, and *Care of Collections*, a reader in preventative conservation.

Ken Arnold wrote his PhD on *Cabinets for the Curious* and was awarded the degree from Princeton University in 1992. Since then he has worked in a variety of museums including the Smithsonian Institute, the Museum of Mankind and the Livesey Museum (Southwark's children's museums) since then. He is now Exhibitions Officer at the Wellcome Institute for the History of Medicine.

Ian Simmons is the Assistant Keeper in charge of hands-on science at Snibston Discovery Park in Coalville, Leicestershire, part of Leicestershire County Council's Museum, Arts and Records Service. He is responsible for all aspects of the creation and development of the museum's hands-on galleries, 'Science Alive' and 'Light Fantastic', as well as the outdoor 'Science Play Area' and the hands-on interpretation of collections. He is a founder of the British Interactive Group and recently completed an MA thesis on setting up hands-on centres within museums for the Department of Museum Studies at Leicester University.

Arthur P. Molella is Head Curator of the 'Science in American Life' exhibition. His previous exhibitions include 'FDR: The Intimate Presidency', an exploration of Roosevelt's revolutionary political uses of the media. Dr Molella is currently Assistant Director for History at the Smithsonian's National Museum of American History and Director of the Museum's newly launched Lemelson Center, a multi-faceted programme for the public understanding of invention, innovation and society. From 1983 to 1994 he was Chairman of the Museums' Department of the History of Science and Technology, where he oversaw the reinstallation of the department's major science and technology exhibitions. From 1981 to 1983 he was curator of electrical technology in the Division of Electricity and Modern Physics of the Museum of American History. A specialist in the history of modern science and technology, Dr Molella received his doctorate in the history of science from Cornell University in 1972. From 1970 to 1981 he served as Associate Editor of the *Papers of Joseph Henry*, the eminent nineteenth-century electrical physicist and first Secretary of the Smithsonian. Dr Molella's other publications include studies of the work of

Sigfried Giedon, Lewis Mumford, and A.P. Usher, pioneering figures in the history of technology. With Carlene Stephens and Robert Kargon, he is currently preparing *Science in American Life*, a book based on the major themes of the exhibition.

Carlene Stephens is a curator in the Department of History, National Museum of American History. She served as deputy chief curator for 'Science in American Life', and in her twenty years at the museum she has worked on numerous other exhibitions, large and small. A historian of technology, her research interests are in the social, cultural and technical history of timekeeping. She has published on the standardization of time the United States, and she is developing a new timekeeping exhibition for the museum, which will open later this decade.

Gillian Thomas joined the development team at the City for Science and Industry, La Villette, Paris, as a science adviser after reading Chemistry at Oxford and an initial period teaching. After its opening, she became responsible for the children's section and the development of activities for young people throughout the Centre. In 1985, she was invited to become the Director of the project to set up Eureka! the Museum for Children, and piloted this through the development and building phase. This opened successfully in July 1992 and Gillian became Head of the new Project Development Division at the Science Museum, London, in January 1993. She is also Vice-President of ECSITE, the European Collaborative for Science, Industry and Technology Exhibitions.

Tim Caulton is lecturer in the Leisure Management Unit at the University of Sheffield, and an independent museum consultant. He was formerly Head of Education and Interpretation at Eureka! The Museum for Children, and Keeper of Extension Services at Sheffield City Museums.

Ibrahim Yahya obtained his first and higher degrees in Physics from the University of Madras in 1981 and 1983 respectively. Formerly an Assistant Professor of Physics and a physicist by training, he entered the museum field in 1986 as Curator of Physics in the Birla Industrial and Technological Museum, National Council of Science Museums, Calcutta. Since then he has planned two new interactive galleries – The Children's Gallery and Vibration Gallery. In 1991 he took study leave and has been studying for a PhD at the Department of Museum Studies, University of Leicester; he is looking at museum learning from the perspective of learning styles. He has published a number of articles on science learning and science centres.

John Durant was born in Norwich in 1950. He attended the City of Norwich Grammar School from 1961 to 1968, and was awarded a Foundation Scholarship in Natural Sciences at Queens' College, Cambridge in 1969. He took a first class honours degree in Zoology in 1972, whereupon he embarked upon a programme of doctoral research in the University of Cambridge, Department of History and Philosophy of Science. He was awarded a PhD for his research on late nineteenth-century evolutionary thought in 1977. From 1976 to 1982 he worked as a staff tutor in Biological Sciences in the University College of Swansea Department for Extra-Mural Studies. From 1983 to 1989 he occupied a similar post in the University of Oxford Department for External Studies. In April 1989, John Durant moved to his present post as Assistant Director at the Science Museum and Head of the Science Communication Division, and Professor of Public Understanding of Science, Imperial College, London. The MSc course in Science Communication at Imperial College was started by Professor Durant in 1991. John Durant has published widely on the history of the life sciences, the relationship between science and human values, and the public understanding of science. He is the co-author of *Aggression: the Myth of Beast Within* (Longman, 1989), and the editor of several volumes, including: *Darwinism and Divinity: Essays on Evolution and Religious Belief* (1985); *Human Origins* (1989) and *Museums and the Public Understanding of Science* (1992). He is the editor of the quarterly international journal, *Public Understanding of Science*. John Durant writes frequently for the popular press, and he contributes from time to time to radio and television science programmes. He presented '4th Dimension', a science magazine programme on Channel 4, and he has made programmes for the BBC's 'Antenna' and 'QED' series. He hosts a BBC Radio 4 science discussion series each summer, under the title 'Science Friction'.

Index